MIX
Papier aus verantwortungsvollen Quellen
Paper from responsible sources
FSC® C105338

FSC
www.fsc.org

Osama Mohammed Elmardi

Solutions to Problems in Heat Transfer

Transient Conduction or Unsteady Conduction

Anchor Academic
Publishing

Mohammed Elmardi, Osama: Solutions to Problems in Heat Transfer. Transient Conduction or Unsteady Conduction, Hamburg, Anchor Academic Publishing 2017

Buch-ISBN: 978-3-96067-123-7
PDF-eBook-ISBN: 978-3-96067-623-2
Druck/Herstellung: Anchor Academic Publishing, Hamburg, 2017

Bibliografische Information der Deutschen Nationalbibliothek:
Die Deutsche Nationalbibliothek verzeichnet diese Publikation in der Deutschen Nationalbibliografie; detaillierte bibliografische Daten sind im Internet über http://dnb.d-nb.de abrufbar.

Bibliographical Information of the German National Library:
The German National Library lists this publication in the German National Bibliography. Detailed bibliographic data can be found at: http://dnb.d-nb.de

All rights reserved. This publication may not be reproduced, stored in a retrieval system or transmitted, in any form or by any means, electronic, mechanical, photocopying, recording or otherwise, without the prior permission of the publishers.

Das Werk einschließlich aller seiner Teile ist urheberrechtlich geschützt. Jede Verwertung außerhalb der Grenzen des Urheberrechtsgesetzes ist ohne Zustimmung des Verlages unzulässig und strafbar. Dies gilt insbesondere für Vervielfältigungen, Übersetzungen, Mikroverfilmungen und die Einspeicherung und Bearbeitung in elektronischen Systemen.

Die Wiedergabe von Gebrauchsnamen, Handelsnamen, Warenbezeichnungen usw. in diesem Werk berechtigt auch ohne besondere Kennzeichnung nicht zu der Annahme, dass solche Namen im Sinne der Warenzeichen- und Markenschutz-Gesetzgebung als frei zu betrachten wären und daher von jedermann benutzt werden dürften.

Die Informationen in diesem Werk wurden mit Sorgfalt erarbeitet. Dennoch können Fehler nicht vollständig ausgeschlossen werden und die Diplomica Verlag GmbH, die Autoren oder Übersetzer übernehmen keine juristische Verantwortung oder irgendeine Haftung für evtl. verbliebene fehlerhafte Angaben und deren Folgen.

Alle Rechte vorbehalten

© Anchor Academic Publishing, Imprint der Diplomica Verlag GmbH
Hermannstal 119k, 22119 Hamburg
http://www.diplomica-verlag.de, Hamburg 2017
Printed in Germany

Dedication

In the name of Allah, the merciful, the compassionate

All praise is due to Allah and blessings and peace is upon his messenger and servant, **Mohammed**, and upon his family and companions and whoever follows his guidance until the day of resurrection.

To the memory of my mother **Khadra Dirar Taha**, my father **Mohammed Elmardi Suleiman**, and my dear aunt **Zaafaran Dirar Taha** who they taught me the greatest value of hard work and encouraged me in all my endeavors.

To my first wife **Nawal Abbas** and my beautiful three daughters **Roa, Rawan** and **Aya** whose love, patience and silence are my shelter whenever it gets hard.

To my second wife **Limya Abdullah** whose love and supplication to Allah were and will always be the momentum that boosts me through the thorny road of research.

To Professor **Mahmoud Yassin Osman** for reviewing and modifying the manuscript before printing process.

This book is dedicated mainly to undergraduate and postgraduate students, especially mechanical and production engineering students where most of the applications are of mechanical engineering nature.

To Mr. **Osama Mahmoud** of Daniya Center for publishing and printing services whose patience in editing and re – editing the manuscript of this book was the momentum that pushed me in completing successfully the present book.

To my friend Professor **Elhassan Mohammed Elhassan Ishag**, Faculty of Medicine, University of Gezira, Medani, Sudan.

To my friend **Mohammed Ahmed Sambo**, Faculty of Engineering and Technology, Nile Valley University, Atbara, Sudan.

To my homeland, Sudan, hoping to contribute in its development and superiority.

Finally, may Allah accept this humble work and I hope that it will be beneficial to its readers

Acknowledgement

I am grateful and deeply indebted to Professor **Mahmoud Yassin Osman** for valuable opinions, consultation and constructive criticism, for without which this work would not have been accomplished.

I am also indebted to published texts in thermodynamics and heat and mass transfer which have been contributed to the author's thinking. Members of Mechanical Engineering Department at Faculty of Engineering and Technology, Nile Valley University, Atbara – Sudan, and Sudan University of Science & Technology, Khartoum – Sudan have served to sharpen and refine the treatment of my topics. The author is extremely grateful to them for constructive criticisms and valuable proposals.

I express my profound gratitude to Mr. **Osama Mahmoud** and Mr. **Ahmed Abulgasim** of Daniya Center for Printing and Publishing services, Atbara, Sudan who they spent several hours in editing, re – editing and correcting the present manuscript.

Special appreciation is due to the British Council's Library for its quick response in ordering the requested bibliography, books, reviews and papers.

Preface

During my long experience in teaching several engineering subjects I noticed that many students find it difficult to learn from classical textbooks which are written as theoretical literature. They tend to read them as one might read a novel and fail to appreciate what is being set out in each section. The result is that the student ends his reading with a glorious feeling of knowing it all and with, in fact, no understanding of the subject whatsoever. To avoid this undesirable end a modern presentation has been adopted for this book. The subject has been presented in the form of solution of comprehensive examples in a step by step form. The example itself should contain three major parts, the first part is concerned with the definition of terms, the second part deals with a systematic derivation of equations to terminate the problem to its final stage, the third part is pertinent to the ability and skill in solving problems in a logical manner.

This book aims to give students of engineering a thorough grounding in the subject of heat transfer. The book is comprehensive in its coverage without sacrificing the necessary theoretical details.

The book is designed as a complete course test in heat transfer for degree courses in mechanical and production engineering and combined studies courses in which heat transfer and related topics are an important part of the curriculum. Students on technician diploma and certificate courses in engineering will also find the book suitable although the content is deeper than they might require.

The entire book has been thoroughly revised and a large number of solved examples and additional unsolved problems have been added. This book contains comprehensive treatment of the subject matter in simple and direct language.

The book comprises eight chapters. All chapters are saturated with much needed text supported and by simple and self-explanatory examples.

Chapter one includes general introduction to transient conduction or unsteady conduction, definition of its fundamental terms, derivation of equations and a wide spectrum of solved examples.

In chapter two the time constant and the response of temperature measuring devices were introduced and discussed thoroughly. This chapter was supported by different solved examples.

Chapter three discusses the importance of transient heat conduction in solids with finite conduction and convective resistances. At the end of this chapter a wide range of solved examples were added. These examples were solved using Heisler charts.

In chapter four transient heat conduction in semi – infinite solids were introduced and explained through the solution of different examples using Gaussian error function in the form of tables and graphs.

Chapter five deals with the periodic variation of surface temperature where the periodic type of heat flow was explained in a neat and regular manner. At the end of this chapter a wide range of solved examples was introduced.

Chapter six concerns with temperature distribution in transient conduction. In using such distribution, the one dimensional transient heat conduction problems could be solved easily as explained in examples.

In chapter seven additional examples in lumped capacitance system or negligible internal resistance theory were solved in a systematic manner, so as to enable the students to understand and digest the subject properly.

Chapter eight which is the last chapter of this book contains unsolved theoretical questions and further problems in lumped capacitance system. How these problems are solved will depend on the full understanding of the previous chapters and the facilities available (e.g. computer, calculator, etc.). In engineering, success depends on the reliability of the results achieved, not on the method of achieving them.

I would like to express my appreciation of the assistance which I have received from my colleagues in the teaching profession. I am particularly indebted to Professor Mahmoud Yassin Osman for his advice on the preparation of this textbook.

When author, printer and publisher have all done their best, some errors may still remain. For these I apologies and I will be glad to receive any correction or constructive criticism.

Assistant Professor/ Osama Mohammed Elmardi Suleiman
Mechanical Engineering Department
Faculty of Engineering and Technology
Nile Valley University, Atbara, Sudan

February 2017

Contents

Dedication i
Acknowledgement ii
Preface iii
Contents vi

Chapter One: Introduction 1
   1.1 General Introduction 2
   1.2 Definition of Lumped Capacity or Capacitance System 3
   1.3 Characteristic Linear Dimensions of Different Geometries 3
   1.4 Derivation of Equations of Lumped Capacitance System 5
   1.5 Solved Examples 8

Chapter Two: Time Constant and Response of Temperature Measuring Instruments 17
   2.1 Introduction 17
   2.2 Solved Examples 18

Chapter Three: Transient Heat Conduction in Solids with Finite Conduction and Convective Resistances 25
   3.1 Introduction 25
   3.2 Solved Examples 27

Chapter Four: Transient Heat conduction in semi – infinite solids 39
   4.1 Introduction 39
   4.2 Penetration Depth and Penetration Time 43
   4.3 Solved Examples 44

Chapter Five: Systems with Periodic Variation of Surface Temperature ... 52
   5.1 Introduction 52
   5.2 Solved Examples 54

Chapter Six: Transient Conduction with Given Temperature Distribution ... 56
   6.1 Introduction 56
   6.2 Solved Examples 56

Chapter Seven: Additional Solved Examples in Lumped Capacitance System.. 59

7.1 Example (1): Determination of Temperature and Rate of Cooling of a Steel Ball 59

7.2 Example (2): Calculation of the Time Required to Cool a Thin Copper Plate.. 60

7.3 Example (3): Determining the Conditions under which the Contact Surface Remains at Constant Temperature 63

7.4 Example (4): Calculation of the Time Required for the Plate to Reach a Given Temperature 64

7.5 Example (5): Determination of the Time Required for the Plate to Reach a Given Temperature 65

7.6 Example (6): Determining the Temperature of a Solid Copper Sphere at a Given Time after the Immersion in a Well – Stirred Fluid 67

7.7 Example (7): Determination of the Heat Transfer Coefficient 68

7.8 Example (8): Determination of the Heat Transfer Coefficient 69

7.9 Example (9): Calculation of the Initial Rate of Cooling of a Steel Ball 70

7.10 Example (10): Determination of the Maximum Speed of a Cylindrical Ingot inside a Furnace 71

7.11 Example (11): Determining the Time Required to Cool a Mild Steel Sphere, the Instantaneous Heat Transfer Rate, and the Total Energy Transfer 72

7.12 Example (12): Estimation of the Time Required to Cool a Decorative Plastic Film on Copper Sphere to a Given Temperature using Lumped Capacitance Theory 75

7.13 Example (13): Calculation of the Time Taken to Boil an Egg 76

7.14 Example (14): Determining the Total Time Required for a Cylindrical Ingot to be heated to a Given Temperature 78

Chapter Eight: Unsolved Theoretical Questions and Further Problems in Lumped Capacitance System 82

8.1 Theoretical Questions 82

8.2 Further Problems 82

References 88

Appendix: Mathematical Formulae Summary 91

Chapter One

Introduction

From the study of thermodynamics, you have learned that energy can be transferred by interactions of a system with its surroundings. These interactions are called work and heat. However, thermodynamics deals with the end states of the process during which an interaction occurs and provides no information concerning the nature of the interaction or the time rate at which it occurs. The objective of this textbook is to extend thermodynamic analysis through the study of transient conduction heat transfer and through the development of relations to calculate different variables of lumped capacitance theory.

In our treatment of conduction in previous studies we have gradually considered more complicated conditions. We began with the simple case of one dimensional, steady state conduction with no internal generation, and we subsequently considered more realistic situations involving multidimensional and generation effects. However, we have not yet considered situations for which conditions change with time.

We now recognize that many heat transfer problems are time dependent. Such unsteady, or transient problems typically arise when the boundary conditions of a system are changed. For example, if the surface temperature of a system is altered, the temperature at each point in the system will also begin to change. The changes will continue to occur until a steady state temperature distribution is reached. Consider a hot metal billet that is removed from a furnace and exposed to a cool air stream. Energy is transferred by convection and radiation from its surface to the surroundings. Energy transfer by conduction also occurs from the interior of the metal to the surface, and the temperature at each point in the billet decreases until a steady state condition is reached. The final properties of the metal will depend significantly on the time – temperature history that results from heat transfer. Controlling the heat transfer is one key to fabricating new materials with enhanced properties.

Our objective in this textbook is to develop procedures for determining the time dependence of the temperature distribution within a solid during a transient process, as well as for determining heat transfer between the solid and its surroundings. The nature of the procedure depends on assumptions that may be made for the process. If, for example, temperature gradients within the solid may be neglected, a comparatively simple approach, termed the lumped capacitance method or negligible internal resistance theory, may be used to determine the variation of temperature with time.

Under conditions for which temperature gradients are not negligible, but heat transfer within the solid is one dimensional, exact solution to the heat equation may be used to compute the dependence of temperature on both location and time. Such solutions are for finite and infinite solids. Also, the response of a semi – infinite solid to periodic heating conditions at its surface is explored.

1.1 General Introduction

Transient conduction is of importance in many engineering aspects for an example, when an engine is started sometime should elapse or pass before steady state is reached. What happens during this lap of time may be detrimental. Again, when quenching a piece of metal, the time history of the temperature should be known (i.e. the temperature – time history). One of the cases to be considered is when the internal or conductive resistance of the body is small and negligible compared to the external or convective resistance.

This system is also called lumped capacity or capacitance system or negligible internal resistance system because internal resistance is small, conductivity is high and the rate of heat flow by conduction is high and therefore, the variation in temperature through the body is negligible. The measure of the internal resistance is done by the Biot (Bi) number which is the ratio of the conductive to the convective resistance.

$$i.e.\ Bi = \frac{hl}{k}$$

When $Bi \ll 0.1$ the system can be assumed to be of lumped capacity (i.e. at $Bi = 0.1$ the error is less than 5% and as Bi becomes less the accuracy increases).

1.2 Definition of Lumped Capacity or Capacitance System

Is the system where the internal or conductive resistance of a body is very small or negligible compared to the external or convective resistance.

Biot number (Bi): is the ratio between the conductive and convective resistance.

$$Bi = \frac{hl}{k} \rightarrow Bi = \frac{conduction\ resistance}{convection\ resistance} = \frac{x}{kA} \Big/ \frac{1}{hA} = \frac{x}{kA} \times \frac{hA}{1} = \frac{hx}{k}$$

Where x is the characteristic linear dimension and can be written as l, h is the heat transfer coefficient by convection and k is the thermal conductivity.
When $Bi \ll 0.1$, the system is assumed to be of lumped capacity.

1.3 Characteristic Linear Dimensions of Different Geometries

The characteristic linear dimension of a body, $L = \frac{V}{A_s} = \frac{volume\ of\ the\ box}{surface\ area\ of\ the\ body}$

Characteristic linear dimension of a plane surface, $L = \frac{t}{2}$

Characteristic linear dimension of a cylinder, $L = \frac{r}{2}$

Characteristic linear dimension of a sphere (ball), $L = \frac{r}{3}$

Characteristic linear dimension of a cube, $L = \frac{a}{6}$

Where: t is the plate thickness, r is the radius of a cylinder or sphere, and a is the length side of a cube.

The derivations of the above characteristic lengths are given below:

i) The characteristic length of plane surface, $L = \frac{t}{2}$

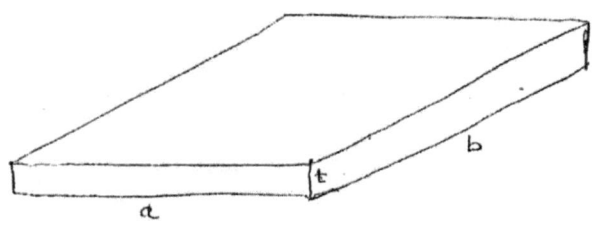

$$V = abt$$

$$A_s = 2at + 2bt + 2ab$$

Since, t is very small; therefore, it can be neglected.

$$A_s = 2ab$$

$$L = \frac{V}{A_s} = \frac{abt}{2ab} = \frac{t}{2}$$

ii) The characteristic length of cylinder,

$$L = \frac{V}{A_s} = \frac{r}{2}$$

$$V = \pi r^2 L$$

$$A_s = 2\pi r L$$

$$\therefore L = \frac{V}{A_s} = \frac{\pi r^2 L}{2\pi r L} = \frac{r}{2}$$

iii) The characteristic length of a sphere (ball),

$$L = \frac{r}{3}$$

$$v = \frac{4}{3}\pi r^3$$

$$A_s = 4\pi r^2$$

$$\therefore L = \frac{V}{A_s} = \frac{\frac{4}{3}\pi r^3}{4\pi r^2} = \frac{r}{3}$$

iv) The characteristic length of a cube,

$$L = \frac{a}{6}$$

$$V = a^3$$

$$A_s = 6a^2$$

$$L = \frac{V}{A_s} = \frac{a^3}{6a^2} = \frac{a}{6}$$

1.4 Derivation of Equations of Lumped Capacitance System

Consider a hot body of an arbitrary shape as shown in Fig. (1.1) below:

Energy balance at any instant requires that:

The rate of loss of internal energy of the body must be equal to the rate of convection from the body to the surrounding fluid.

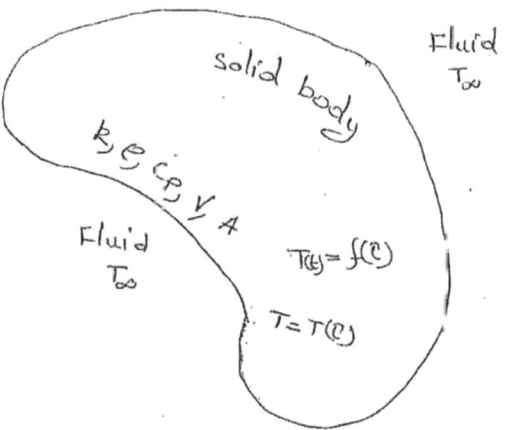

Fig. (1.1) Hot body of an arbitrary shape

$$q = -\rho V c_p \frac{dT(t)}{d\tau} = hA_s(T(t) - T_\infty) \quad (1.1)$$

Put $(T(t) - T_\infty) = \theta$

And,

$$\frac{dT(t)}{d\tau} = \frac{d\theta}{d\tau}$$

$$\therefore -\rho V c_P \frac{d\theta}{d\tau} = h A_s \theta \qquad (1.2)$$

If the temperature of the body at time $\tau = 0$ is equal to T_o

$$\therefore \theta_o = T_o - T_\infty$$

$$-\rho V c_P \frac{d\theta}{\theta} = h A_s \, d\tau \qquad (1.3)$$

By integrating the above equation (1.3) we obtain the following equation:

$$-\rho V c_P \int_{\theta_o}^{\theta} \frac{d\theta}{\theta} = \int_{\tau=0}^{\tau=\tau} h A_s \, d\tau$$

$$-\rho V \ln \frac{\theta}{\theta_o} = h A_s \, \tau$$

$$\ln \frac{\theta}{\theta_o} = -\frac{h A_s \tau}{\rho V c_P}$$

$$\therefore \frac{\theta}{\theta_o} = e^{\frac{-h A_s \tau}{\rho V c_P}} \qquad (1.4)$$

$$\frac{h A_s}{\rho V c_P} \cdot \tau \text{ can be written as } \frac{h V}{k A_s} \cdot \frac{A_s^2 k}{V^2 \rho c_P} \tau \qquad (1.5)$$

$$\frac{k}{\rho c_P L^2} \cdot \tau = \text{Fourier number Fo (dimensionless Quantity)} \qquad (1.6)$$

And,

$$\frac{h L}{k} = Bi \qquad (1.7)$$

$$\therefore \frac{h A_s}{\rho V c_P} \cdot \tau = Bi \times Fo \qquad (1.8)$$

$$\therefore \frac{\theta}{\theta_o} = \frac{T(t) - T_\infty}{T_o - T_\infty} = e^{-Bi \times Fo} \qquad (1.9)$$

The instantaneous heat transfer rate $q'(\tau)$ is given by the following equation:

$$q'(\tau) = hA_s\theta = hA_s\theta_o e^{-Bi \times Fo} \quad (1.10)$$

At time $\tau = 0$,

$$q'(\tau) = hA_s\theta_o \quad (1.11)$$

The total heat transfer rate from $\tau = 0$ to $\tau = \tau$ is given by the following equation:

$$Q(t) = \int_{\tau=0}^{\tau=\tau} q'(\tau) = \int_0^\tau hA_s\theta_o e^{-Bi \times Fo} \quad (1.12)$$

From equation (1.8) $\rightarrow Bi \times Fo = \dfrac{hA_s}{\rho V c_p} \cdot \tau$ and substitute in equation (1.12), the following equation is obtained:

$$Q(t) = hA_s\theta_o \int_0^\tau e^{\frac{-hA_s}{\rho V c_p}\cdot\tau} = hA_s\theta_o \left[\dfrac{e^{\frac{-hA_s}{\rho V c_p}\cdot\tau}}{\frac{-hA_s}{\rho V c_p}}\right]_0$$

$$= hA_s\theta_o \, e^{\frac{-hA_s}{\rho V c_p}\cdot\tau} \times \dfrac{-\rho V c_p}{hA_s} = -hA_s\theta_o \dfrac{\rho V c_F}{hA_s}\left[e^{\frac{-hA_s}{\rho V c_p}\cdot\tau}\right]_0^\tau$$

$$= -hA_s\theta_o \dfrac{\rho V c_p}{hA_s} e^{\frac{-hA_s}{\rho V c_p}\cdot\tau} - \left\{-hA_s\theta_o \dfrac{\rho V c_p}{hA_s}\right\}$$

$$= -hA_s\theta_o \dfrac{\rho V c_p}{hA_s} e^{\frac{-hA_s}{\rho V c_p}\cdot\tau} + hA_s\theta_o \dfrac{\rho V c_p}{hA_s}$$

$$= hA_s\theta_o \dfrac{\rho V c_p}{hA_s} \left\{1 - e^{\frac{-hA_s}{\rho V c_p}\cdot\tau}\right\}$$

$$= hA_s\theta_o \cdot \dfrac{\tau}{Bi \times Fo}\{1 - e^{-Bi \times Fo}\} \quad (1.13)$$

1.5 Solved Examples

Example (1):

Chromium steel ball bearing ($k = 50 w/mK, \alpha = 1.3 \times 10^{-5} m^2/s$) are to be heat treated. They are heated to a temperature of $650°C$ and then quenched in oil that is at a temperature of $55°C$.

The ball bearings have a diameter of 4cm and the convective heat transfer coefficient between the bearings and oil is $300 w/m^2 K$. Determine the following:

(a) The length of time that the bearings must remain in oil before the temperature drops to $200°C$.

(b) The total heat removed from each bearing during this time interval.

(c) Instantaneous heat transfer rate from the bearings when they are first placed in the oil and when they reach $200°C$.

Solution:

Chromium steel ball bearings

$k = 50 w/mK$

$\alpha = thermal\ diffusivity = 1.3 \times 10^{-5} m^2/s$

$T_o = 650°C, \quad T_\infty = 55°C$

Diameter of ball bearing, $d = 4cm = 0.04m$

$h = 300 w/m^2 K$

a) $\tau = ?\quad T(t) = 200°C$

$$Bi = \frac{hL}{k}$$

Characteristic length, $L = \dfrac{\text{volume of the ball}}{\text{surface area of the ball}} = \dfrac{V}{A_s}$

$$\therefore L = \frac{v}{A_s} = \frac{\frac{4}{3}\pi r^3}{4\pi r^2} = \frac{r}{3}$$

$$Bi = \frac{hL}{k} = \frac{300 \times 0.04}{6 \times 50} = 0.04$$

Since $Bi \ll 0.1$, the lumped capacity system or the negligible internal resistance theory is valid.

$$\frac{\theta}{\theta_o} = \frac{T(t) - T_\infty}{T_o - T_\infty} = e^{-Bi\,Fo}$$

$$Fo = \frac{k}{\rho c_p L^2} \cdot \tau = \frac{\alpha}{L^2} \cdot \tau = \frac{1.3 \times 10^{-5}}{\left(\frac{0.02}{3}\right)^2} = 0.2925\,\tau$$

$$\frac{\theta}{\theta_o} = \frac{200 - 55}{650 - 55} = e^{-0.04 \times 0.2925\,\tau}$$

$$0.2437 = e^{-0.0117\,\tau}$$

$$\log 0.2437 = \log e^{-0.0117\,\tau} = -0.0117\,\tau \log e$$

$$\therefore \tau = \frac{\log 0.2437}{\log e \times -0.0117} = \frac{\log 0.2437}{-0.0117 \log e} = 120.7\,s$$

b) Q (t) =?

$$Q(t) = hA_s\theta_o[1 - e^{-Bi \times Fo}]\frac{\tau}{Bi \times Fo}$$

$$Q(t) = 300 \times 4\pi \times 0.02^2 (650 - 55)[1 - e^{-0.04 \times 0.2925 \times 120.7}]\frac{120.7}{0.04 \times 0.2925 \times 120.7}$$

$$Q(t) = 58005.4\,J \simeq 58\,kj$$

c) Instantaneous heat transfer rate, $q'(\tau) = ?$

i) When they are first placed in oil,

$$q'(\tau) = hA_s\theta_o = 300 \times 4\pi \times 0.02^2 (650 - 55) = 897.24w$$

ii) When they reach $200°c$,

$$q'(\tau) = hA_s\theta_o e^{-Bi \times Fo} = 897.24 \times e^{-0.04 \times 0.2925 \times 120.7} = 218.6w$$

Example (2):

The product from a chemical process is in the form of pellets which are approximately spherical to the mean diameter, $d = 4mm$. These pellets are initially at 403K and must be cooled to 343K maximum before entering storage vessel. This proposed to cool these pellets to the required temperature by passing them down slightly inclined channel where they are subjected to a stream of air at 323K. If the length of the channel is limited to 3m, calculate the maximum velocity of the pellets along the channel and the total heat transferred from one pellet.

Heat transfer from pellet surface to the air stream may be considered to be the limiting process with $\frac{h\,d}{k_a} = 2$.

Where:

h = heat transfer coefficient at the pellet surface.
k_a = thermal conductivity of air = $0.13 w/mK$

Other data:

Pellet material density = $480 kg/m^3$
Specific heat capacity $c_P = 2 kj/kgK$
You may assume that the lumped capacitance system theory is applicable.

Solution:

The product from a chemical process in the form of spherical pellets,
$d = 4mm = 0.004m, \quad r = 0.002m$
$T_o = 403\ K$
$T(t) = 343\ K$
$T_\infty = 323\ K$
Length of the channel, $L = 3m$

Calculate:

* The maximum velocity of the pellets along the channel, $v_{max} = ?$
* $Q(t) = ?$

Heat transfer from pellet surface to the air stream is limited to $\frac{h\,d}{k_a} = 2$

$k_a = 0.13 w/mK$

ρ pellet $= 480 kg/m^3$

$c_p = 2 kj/kgK = 2 \times 10^3 j/kgK$

It is assumed that lumped capacity system or negligible internal resistance theory is applicable.

Max. Velocity, $v_{max} = \dfrac{L}{\tau}$

Biot Number, $Bi = \dfrac{hL}{k}$

Characteristic length, L,

$$L = \dfrac{\text{volume of a sphere}}{\text{surface area of a sphere}} = \dfrac{V}{A_s} = \dfrac{\frac{4}{3}\pi r^3}{4\pi r^2} = \dfrac{r}{3}$$

$$\therefore L = \dfrac{r}{3} = \dfrac{0.002}{3} m$$

$$Bi = \dfrac{hr}{3k} = \dfrac{0.002h}{3k}$$

$$\dfrac{hd}{k_a} = 2 \;\; ; \;\; \dfrac{h \times 0.004}{0.13} = 2$$

$$\therefore h = \dfrac{2 \times 0.13}{0.004} = 65 w/m^2 K$$

$$Bi = \dfrac{0.002 \times 65}{3k} = \dfrac{0.13}{3k}$$

$$\dfrac{\theta}{\theta_o} = \dfrac{T(t) - T_\infty}{T_o - T_\infty} = e^{-Bi \times Fo}$$

$$\dfrac{\theta}{\theta_o} = \dfrac{343 - 323}{403 - 323} = e^{-\frac{0.13}{3k} \times Fo}$$

$$Fo = \dfrac{k}{\rho c_p L^2} \cdot \tau = \dfrac{k}{480 \times 2 \times 10^3 \times \left(\dfrac{0.002}{3}\right)^2} \cdot \tau$$

$$Fo = 2.34375 k \tau$$

$$\frac{\theta}{\theta_o} = 0.25 = e^{-\frac{0.13}{3k} \times 2.34375k\,\tau}$$

$$0.25 = e^{-0.1015625\,\tau}$$

$$\log 0.25 = -0.1015625\,\tau \log e$$

$$\therefore \tau = \frac{\log 0.25}{\log e \times -0.1015625} = \frac{\log 0.25}{-0.1015625 \log e} = 13.65\ s$$

$$\therefore \text{max. velocity}, v_{max} = \frac{L}{\tau} = \frac{3}{13.65} = 0.22 m/s$$

$$Q(t) = hA_s\theta_o\left[1 - e^{-Bi \times Fo}\right]\frac{\tau}{Bi \times Fo}$$

$$Q(t) = 65 \times 4\pi \times 0.002^2(403 - 323)\left[1 - e^{-\frac{0.13}{3k} \times 2.34375k \times 13.65}\right] \times \frac{13.65}{\frac{0.13}{3} \times 2.34375 \times 13.05} = 1.93\ J/\text{pellet}$$

$$\therefore Q(t) = 1.93\ J/\text{pellet}$$

Example (3):

A piece of chromium steel of length 7.4cm (density=8780kg/m³ ; $k = 50w/mK$ and specific heat $c_P = 440j/kgK$) with mass 1.27kg is rolled into a solid cylinder and heated to a temperature of 600°C and quenched in oil at 36°c. Show that the lumped capacitance system analysis is applicable and find the temperature of the cylinder after 4min. What is the total heat transfer during this period?
You may take the convective heat transfer coefficient between the oil and cylinder at 280w/m²k.

Solution:

A piece of chromium steel, L = 7.4cm = 0.074m

$\rho = 8780 kg/m^3$; $k = 50w/mK$; $c_p = 440j/kgK$; $m = 1.27kg$

Rolled into a solid cylinder,

$T_o = 600°C$; $T_\infty = 36°C$; $h = 280 w/m^2 K$

$T(t) = ?$ after $4 min$. (i.e. $\tau = 4 \times 60 = 240s$); $Q(t) = ?$

$$Bi = \frac{hL}{k}$$

Characteristic length, $L = \frac{\text{volume of a cylinder}}{\text{surface area of a cylinder}} = \frac{V}{A_s}$

$$= \frac{\pi r^2 L}{2\pi r L} = \frac{r}{2}$$

$$\therefore Bi = \frac{hr}{2k}$$

$$V = \frac{m}{\rho} = \frac{1.27}{3780} = 1.4465 \times 10^{-4} m^3$$

$$L = 7.4cm = 0.074m$$

$$V = \pi r^2 L = 1.4465 \times 10^{-4}$$

$$\therefore r = \sqrt{\frac{1.4465 \times 10^{-4}}{\pi \times 0.074}} = 0.025m$$

$$Bi = \frac{hr}{2\kappa} = \frac{280 \times 0.025}{2 \times 50} = 0.07$$

Since $Bi \ll 0.1$, then the system is assumed to be of lumped capacitance system, and therefore the lumped capacitance system analysis or the negligible internal resistance theory is applicable.

$T(t) = ?$

$\tau = 4 \times 60 = 240 \ s$

$$\frac{\theta}{\theta_o} = \frac{T(t) - T_\infty}{T_o - T_\infty} = e^{-Bi \, Fo}$$

$$Fo = \frac{k}{\rho c_p L^2} \cdot \tau$$

$$Fo = \frac{50}{8780 \times 440 \times \left(\frac{0.025}{2}\right)^2} \times 240 = 19.88$$

$$\frac{\theta}{\theta_o} = \frac{T(t) - 36}{600 - 36} = e^{-0.07 \times 19.88}$$

$$\frac{T(t) - 36}{564} = e^{-1.3916}$$

$$\therefore T(t) = 56 + e^{-1.3916} + 36 = 176.254°C$$

$$q'(\tau)^{=0} = q'(0) = hA_s\theta_o = hA_s(T_o - T_\infty)$$

$$= 280 \times 2\pi \times 0.025 \times 0.074(600 - 36) = 1835.65w$$

$$q'(\tau)^{=4min} = hA_s\theta_o e^{-Bi \times Fo}$$

$$= 1835.65 \times e^{-1.3916} = 456.5w$$

Total heat transfer rate,

$$Q(t) = hA_s\theta_o(1 - e^{-Bi \times Fo}) \frac{\tau}{Bi \times Fo}$$

$$Q(t) = 1835.65(1 - e^{-1.3916}) \frac{240}{1.3916} = 237855.57 \, J$$

$$\simeq 237.86 \, kj$$

Example (4):

A piece of aluminum ($\rho = 2705 kg/m^3$, $k = 216 w/mK$, $c_p = 896 j/kgK$) having a mass of 4.78kg and initially at temperature of $290°C$ is suddenly immersed in a fluid at $15°C$.

The convection heat transfer coefficient is $54 w/m^2K$. Taking the aluminum as a sphere having the same mass as that given, estimate the time required to cool the aluminum to $90°C$.

Find also the total heat transferred during this period. (Justify your use of the lumped capacity method of analysis).

Solution:

A piece of aluminum

$\rho = 2705 kg/m^3; \quad k = 216 w/mK; \quad c_p = 896 J/kgK; \quad m = 4.78 kg$

$T_o = 290°C;$

$T_\infty = 15°C;$

$h = 24 w/m^2 K;$

Taking aluminum as a sphere.

Estimate $\tau = ?$ $T(t) = 90°C;$ $Q(t) = ?$

$$\frac{\theta}{\theta_o} = \frac{T(t) - T_\infty}{T_o - T_\infty} = e^{-Bi \times Fo}$$

$$Bi = \frac{hL}{k}$$

The characteristic length, $L = \frac{V}{A_s} = \frac{\frac{4}{3}\pi r^3}{4\pi r^2} = \frac{r}{3}$

$$\therefore Bi = \frac{hr}{3k}$$

$$\rho = \frac{m}{V}; \quad V = \frac{m}{\rho} = \frac{4.78}{2705} = \frac{4}{3}\pi r^3$$

$$r = \sqrt[3]{\frac{4.78}{2705} \times \frac{3}{4\pi}} = 0.075 m$$

$$\therefore Bi = \frac{54 \times 0.075}{3 \times 216} = 0.00625$$

If $Bi \ll 0.1$, then the system is assumed to be of lumped capacity.

Since $Bi = 0.00625 \ll 0.1$, therefore the lumped capacity method of analysis is used.

$$Fo = \frac{k}{\rho c_p L^2} \cdot \tau = \frac{216}{2705 \times 896 \times \left(\frac{0.075}{3}\right)^2} \cdot \tau$$

$$Fo = 0.1426\, \tau$$

$$\frac{\theta}{\theta_o} = \frac{90-15}{290-15} = e^{-0.00625 \times 0.1426\, \tau}$$

$$\frac{75}{275} = e^{-8.9125 \times 10^{-4}\, \tau}$$

$$\log\frac{75}{275} = -8.9125 \times 10^{-4}\, \tau \log e$$

$$\therefore \tau = \frac{\log\frac{75}{275}}{-8.9125 \times 10^{-4}\, \tau \log e} = 1457.8\, s$$

Total heat transfer rate, Q(t),

$$Q(t) = hA_s\theta_o\left[1 - e^{-Bi \times Fo}\right]\frac{\tau}{Bi \times Fo}$$

$$Q(t) = 54 \times 4\pi \times 0.075^2 (290-15)\left[1 - e^{-8.9125 \times 10^{-4} \times 1457.8}\right] \times$$

$$\frac{1457.8}{8.9125 \times 10^{-4} \times 1457.8} = 856552\, J = 856.6\, kj$$

$$\therefore\ Q(t) = 856552\, J = 856.6\, kj$$

Chapter Two

Time Constant and Response of Temperature Measuring Instruments

2.1 Introduction

Measurement of temperature by a thermocouple is an important application of the lumped parameter analysis. The response of a thermocouple is defined as the time required for the thermocouple to attain the source temperature.

It is evident from equation (1.4), that the larger the quantity $\frac{hA_s}{\rho V c_p}$, the faster the exponential term will approach zero or the more rapid will be the response of the temperature measuring device. This can be accomplished either by increasing the value of "h" or by decreasing the wire diameter, density and specific heat. Hence, a very thin wire is recommended for use in thermocouples to ensure a rapid response (especially when the thermocouples are employed for measuring transient temperatures).

From equation (1.8);

$$\frac{Bi \times Fo}{\tau} = \frac{hA_s}{\rho V c_p}$$

$$\frac{\rho V c_p}{hA_s} = \frac{\tau}{Bi \times Fo}$$

The quantity $\frac{\rho V c_p}{hA_s}$ (which has units of time) is called time constant and is denoted by the symbol τ^*. Thus,

$$\frac{\tau}{Bi \times Fo} = \tau^* = \frac{\rho V c_p}{hA_s} = \frac{k}{\alpha h} \cdot \frac{V}{A_s} \qquad (2.1)$$

$$\left\{ \text{Since } \alpha = \frac{k}{\rho c_p} \right\}$$

And,

$$\frac{\theta}{\theta_o} = \frac{T(t) - T_\infty}{T_o - T_\infty} = e^{-Bi \times Fo} = e^{-(\tau/\tau^*)} \quad (2.2)$$

At $\tau = \tau^*$ (one time constant), we have from equation (2.2),

$$\frac{\theta}{\theta_o} = \frac{T(t) - T_\infty}{T_o - T_\infty} = e^{-1} = 0.368 \quad (2.3)$$

Thus, τ^* is the time required for the temperature change to reach 36.8% of its final value in response to a step change in temperature. In other words, temperature difference would be reduced by 63.2%. The time required by a thermocouple to reach its 63.2% of the value of initial temperature difference is called its sensitivity. Depending upon the type of fluid used the response times for different sizes of thermocouple wires usually vary between 0.04 to 2.5 seconds.

2.2 Solved Examples

Example (1):

A thermocouple junction of spherical form is to be used to measure the temperature of a gas stream.

$h = 400 w/m^{2 o}C$; k(thermocouple junction) $= 20 w/m^o C$; $c_p = 400 J/kg^o C$;
and $\rho = 8500 kg/m^3$;

Calculate the following:

(i) Junction diameter needed for the thermocouple to have thermal time constant of one second.

(ii) Time required for the thermocouple junction to reach $198^o c$ if junction is initially at $25^o C$ and is placed in gas stream which is at $200^o C$.

Solution:

Given: $h = 400 w/m^{2 o}C$; k(thermocouple junction) $= 20 w/m^o C$;

$c_p = 400 J/kg^o C$; $\rho = \dfrac{8500 kg}{m^3}$.

(i) Junction diameter, d =?

τ^* (thermal time constant) = 1 s

The time constant is given by:

$$\tau^* = \frac{\rho V c_P}{h A_s} = \frac{\rho \times \frac{4}{3}\pi r^3 \times c_P}{h \times 4\pi r^2} = \frac{\rho r c_P}{3h}$$

$$\text{or } 1 = \frac{8500 \times r \times 400}{3 \times 400}$$

$$\therefore r = \frac{3}{8500} = 3.53 \times 10^{-4} m = 0.353 \, mm$$

$$\therefore d = 2r = 2 \times 0.353 = 0.706 \, mm$$

(ii) Time required for the thermocouple junction to reach $198°C$; $\tau =?$

Given: $T_o = 25°C$; $T_\infty = 200°C$; $T(t) = 198°C$;

$$Bi = \frac{h L_c}{k}$$

L_c of a sphere $= \frac{r}{3}$

$$\therefore Bi = \frac{h(r/3)}{k} = \frac{400 \times (0.353 \times 10^{-3}/3)}{20} = 0.00235$$

As Bi is much less than 0.1, the lumped capacitance method can be used. Now,

$$\frac{T(t) - T_\infty}{T_o - T_\infty} = e^{-Bi \times Fo} = e^{-(\tau/\tau^*)}$$

$$Fo = \frac{k}{\rho c_P L^2} \cdot \tau = \frac{20}{8500 \times 400 \times \left(\frac{0.353 \times 10^{-3}}{3}\right)^2} \cdot \tau = 424.86 \, \tau$$

$$Bi \times Fo = 0.00235 \times 424.86 \, \tau = 0.998\tau$$

$$\therefore Bi \times Fo = \frac{\tau}{\tau^*} = 0.998\tau$$

$\because \tau^* = 1$, therefore, $\tau = 0.998\tau$

$$\therefore \frac{\theta}{\theta_o} = \frac{198-200}{25-200} = e^{-0.998\tau}$$

$$0.01143 = e^{-0.998\tau}$$

$$-0.998\tau \ln e = \ln 0.01143$$

$$\therefore \tau = \frac{\ln 0.01143}{-0.998 \times 1} = 4.48 \ s$$

Example (2):

A thermocouple junction is in the form of 8mm diameter sphere. Properties of material are:

$c_p = 420 J/kg°C$; $\rho = 8000 kg/m^3$; $k = 40 w/m°C$; $h = 40 w/m^{2°}C$;

This junction is initially at $40°C$ and inserted in a stream of hot air at $300°C$. Find the following:

(i) Time constant of the thermocouple.

(ii) The thermocouple is taken out from the hot air after 10 seconds and kept in still air at $30°C$. Assuming the heat transfer coefficient in air is $10 w/m^{2°}C$, find the temperature attained by the junction 20 seconds after removal from hot air.

Solution:

Given: $r = \frac{8}{2} = 4mm = 0.004m$; $c_p = \frac{420 J}{kg°c}$; $\rho = \frac{8000 kg}{m^3}$; $k = 40 w/m°C$

$h = 40 w/m^{2°}C$ (gas stream or hot air); $h = 10 w/m^{2°}C$ (still air)

(i) Time constant of the thermocouple, $\tau^* = ?$

$$\tau^* = \frac{\tau}{Bi \times Fo} = \frac{\rho V c_p}{hA_s} = \frac{\rho \times \frac{4}{3}\pi r^3 \times c_p}{h \times 4\pi r^2} = \frac{\rho r c_p}{3h}$$

$$\therefore \tau^* = \frac{8000 \times 0.004 \times 420}{3 \times 40} = 112 \ s \ \text{(when thermocouple is in gas stream)}$$

(ii) The temperature attained by the junction, ; $T(t) = ?$

Given: $T_o = 40°C$; $T_\infty = 300°C$; $\tau = 10 \ s$;

The temperature variation with respect to time during heating (when dipped in gas stream) is given by:

$$\frac{\theta}{\theta_o} = \frac{T(t) - T_\infty}{T_o - T_\infty} = e^{-Bi \times Fo} = e^{-(\tau/\tau^*)}$$

or $\dfrac{T(t) - 300}{40 - 300} = e^{-(\tau/\tau^*)} = e^{-(10/112)} = 0.9146$

$\therefore T(t) = -260 \times 0.9146 + 300 = 62.2°C$

The temperature variation with respect to time during cooling (when exposed to air) is given by:

$$\frac{T(t) - T_\infty}{T_o - T_\infty} = e^{-(\tau/\tau^*)}$$

where $\tau^* = \dfrac{\rho r c_p}{3h} = \dfrac{8000 \times 0.004 \times 420}{3 \times 10} = 448 \ s$

$\therefore \dfrac{T(t) - 30}{62.2 - 30} = e^{(-20/448)} = 0.9563$

or $T(T) = (62.2 - 30) \times 0.9563 + 30 = 50.79°C$

Example (3):

A very thin glass walled 3mm diameter mercury thermometer is placed in a stream of air, where heat transfer coefficient is $55 w/m^2 °C$, for measuring the unsteady temperature of air. Consider cylindrical thermometer bulb to consist of mercury only for which $k = 8.8 w/m°C$ and $\alpha = 0.0166 m^2/h$. Calculate the time required for the temperature change to reach half its final value.

Solution:

Given: $r = \dfrac{3}{2} = 1.5 mm = 0.0015 m$; $h = 55 w/m^2°C$; $k = 8.8 w/m°C$;
$\alpha = 0.0166 m^2/h$

The time constant is given by:

$$\tau^* = \frac{k}{\alpha h} \cdot \frac{V}{A_s} \quad \text{from equation (2.1)}$$

$$\therefore \tau^* = \frac{k}{\alpha h} \times \frac{\pi r^2 L}{2\pi r L} = \frac{kr}{2\alpha h} = \frac{8.8 \times 0.0015}{2 \times 0.0166 \times 55} = 0.0027229 h = 26 s$$

For temperature change to reach half its final value

$$\frac{\theta}{\theta_o} = \frac{1}{2} = e^{-(\tau/\tau^*)}$$

$$\ln\frac{1}{2} = \ln e^{-\tau/\tau^*}$$

$$\ln\frac{1}{2} = -\tau/\tau^* \ln e$$

$$-\frac{\tau}{\tau^*} = \frac{\ln\frac{1}{2}}{\ln e} = \ln\frac{1}{2} = -0.693$$

$$\therefore \frac{\tau}{\tau^*} = 0.693$$

$$or\ \tau = \tau^* \times 0.693 = 26 \times 0.693 = 18.02\ s$$

Note: Thus, one can expect thermometer to record the temperature trend accurately only for unsteady temperature changes which are slower.

Example (4):

The temperature of an air stream flowing with a velocity of 3m/s is measured by a copper – constantan thermocouple which may be approximately as a sphere of 2.5mm in diameter. Initially the junction and air are at a temperature of 25^oC. The air temperature suddenly changes to and is maintained at 215^oC.

(i) Determine the time required for the thermocouple to indicate a temperature of 165^oC. Also, determine the thermal time constant and the temperature indicated by the thermocouple at that instant.

(ii) Discuss the stability of this thermocouple to measure unsteady state temperature of a fluid when the temperature variation in the fluid has a time period of 3.6s.

The thermal junction properties are:

$\rho = 8750 kg/m^3$; $c_p = 380 J/kg\,°C$; $k(\text{thermocouple}) = 28 w/m\,°C$; and $h = 145 w/m^{2°}C$;

Solution:

Given: $r = \frac{2.5}{2} = 1.25mm = 0.00125m$; $T_o = 25°C$; $T_\infty = 215°C$; $T(t) = 165°C$

Time required to indicate temperature of $165°C$; $\tau = ?$ and $\tau^* = ?$;

Characteristic length,

$$L_c = \frac{V}{A_s} = \frac{\frac{4}{3}\pi r^3}{4\pi r^2} = \frac{r}{3} = \frac{0.00125}{3} = 0.0004167m$$

Thermal diffusivity,

$$\alpha = \frac{k}{\rho c_p} = \frac{28}{8750 \times 380} = 8.421 \times 10^{-6} m^2/s$$

Fourier number,

$$Fo = \frac{k}{\rho c_p L_c^2} \cdot \tau = \frac{\alpha \tau}{L_c^2} = \frac{8.421 \times 10^{-6} \tau}{(0.0004167)^2} = 48.497\,\tau$$

Biot number,

$$Bi = \frac{hL_c}{k} = \frac{145 \times 0.0004167}{28} = 0.002158$$

As $Bi \ll 0.1$, hence lumped capacitance method may be used for the solution of the problem.

The temperature distribution is given by:

$$\frac{\theta}{\theta_o} = \frac{T(t) - T_\infty}{T_o - T_\infty} = e^{-Bi \times Fo}$$

or $\frac{165 - 215}{25 - 215} = e^{(-0.002158 \times 48.497\tau)} = e^{-0.1046\tau}$

$$0.263 = e^{-0.1046\tau}$$

$$-0.1046\tau \ln e = \ln 0.263$$

$$\therefore \tau = \frac{\ln 0.263}{-0.1046 \times 1} = 12.76 \, s$$

Thus, the thermocouple requires 12.76 s to indicate a temperature of $165°C$. The actual time requirement will, however, be greater because of radiation from the probe and conduction along the thermocouple lead wires.

The time constant (τ^*) is defined as the time required to yield a value of unity for the exponent term in the transient relation.

$$Bi \times Fo = \frac{\tau}{\tau^*} = 1$$

Or

$$0.002158 \times 48.497 \, \tau^* = 1$$

Or

$$\tau^* = 9.55 \, s$$

At 9.55 s, the temperature indicated by the thermocouple is given by:

$$\frac{T(t) - T_\infty}{T_o - T_\infty} = e^{-1}$$

Or

$$\frac{T(t) - 215}{25 - 215} = e^{-1}$$

Or

$$T(t) = 215 + (25 - 215)e^{-1} = 145°c$$

(ii) As the thermal time constant is 9.55 s and time required to effect the temperature variation is 3.6 s which is less than the thermal time constant, hence, the temperature recovered by the thermocouple may not be reliable.

Chapter Three

Transient Heat Conduction in Solids with Finite Conduction and Convective Resistances $[0 < Bi < 100]$

3.1 Introduction

As shown in Fig. (3.1) below, consider the heating and cooling of a plane wall having a thickness of 2L and extending to infinity in y and z directions.

Let us assume that the wall, initially, is at uniform temperature T_o and both the surfaces $(x = \pm L)$ are suddenly exposed to and maintained at the ambient (i.e. surrounding) temperature T_∞. The governing differential equation is:

$$\frac{d^2 t}{dx^2} = \frac{1}{\alpha}\frac{dt}{d\tau} \qquad (3.1)$$

The boundary conditions are:

(i) At $\tau = 0$, $T(t) = T_0$

(ii) At $\tau = 0$, $\frac{dT(t)}{dx} = 0$

(iii) At $x = \pm L$; $kA\frac{dT(t)}{dx} = hA(T(t) - T_\infty)$

(The conduction heat transfer equals convective heat transfer at the wall surface).

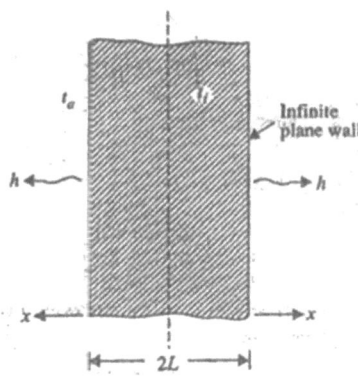

Fig. (3.1) Transient heat conduction in an infinite plane wall

The solutions obtained after rigorous mathematical analysis indicate that:

$$\frac{T(t) - T_\infty}{T_o - T_\infty} = f\left[\frac{x}{L}, \frac{hl}{k}, \frac{\alpha\tau}{l^2}\right] \quad (3.2)$$

From equation (3.2), it is evident that when conduction resistance is not negligible, the temperature history becomes a function of Biot numbers $\left\{\frac{hl}{k}\right\}$, Fourier number $\left\{\frac{\alpha\tau}{l^2}\right\}$ and the dimensionless parameter $\left\{\frac{x}{L}\right\}$ which indicates the location of point within the plate where temperature is to be obtained. The dimensionless parameter $\left\{\frac{x}{L}\right\}$ is replaced by $\left\{\frac{r}{R}\right\}$ in case of cylinders and spheres.

For the equation (3.2) graphical charts have been prepared in a variety of forms. In the Figs. from (3.2) to (3.4) the Heisler charts are shown which depict the dimensionless temperature $\left[\frac{T_c - T_\infty}{T_o - T_\infty}\right]$ versus Fo (Fourier number) for various values of $\left(\frac{1}{Bi}\right)$ for solids of different geometrical shapes such as plates, cylinders and spheres. These charts provide the temperature history of the solid at its mid – plane ($x = 0$) and the temperatures at other locations are worked out by multiplying the mid – plane temperature by correction factors read from charts given in figs. (3.5) to (3.7). The following relationship is used:

$$\frac{\theta}{\theta_o} = \frac{T(t) - T_\infty}{T_o - T_\infty} = \left[\frac{T_c - T_\infty}{T_o - T_\infty}\right] \times \left[\frac{T(t) - T_\infty}{T_c - T_\infty}\right]$$

The values Bi (Biot number) and Fo (Fourier number), as used in Heisler charts, are evaluated on the basis of a characteristic parameter s which is the semi – thickness in the case of plates and the surface radius in case of cylinders and spheres.

When both conduction and convection resistances are almost of equal importance the Heister charts are extensively used to determine the temperature distribution.

3.2 Solved Examples

Example (1):

A 60 mm thickness large steel plate ($k = 42.6 w/m°C, \alpha = 0.043 m^2/h$), initially at $440°C$ is suddenly exposed on both sides to an environment with convective heat transfer coefficient $235 w/m^{2°}C$, and temperature $50°C$. Determine the center line temperature, and temperature inside the plate $15mm$ from the mid – plane after 4.3 minutes.

Solution:

Given: $2L = 60mm = 0.06m, k = 42.6w/m°C, \alpha = 0.043m^2/h, T_o = 440°C,$
$h = 235w/m^{2°}C, T_\infty = 50°C, \theta = 4.3 minutes.$

Temperature at the mid – plane (centerline) of the plate T_c:

The characteristic length, $L_c = \frac{60}{2} = 30mm = 0.03m$

Fourier number, $F_o = \frac{\alpha\tau}{L_c^2} = \frac{0.043 \times (4.3/60)}{(0.03)^2} = 3.424$

Biot number, $Bi = \frac{hL_c}{k} = \frac{235 \times 0.03}{42.6} = 0.165$

At $Bi > 0.1$, the internal temperature gradients are not small, therefore, internal resistance cannot be neglected. Thus, the plate cannot be considered as a lumped system. Further, as the $Bi < 100$, Heisler charts can be used to find the solution of the problem.

Corresponding to the following parametric values, from Heisler charts Fig. (3.2), we have $F_o = 3.424; \frac{1}{Bi} = \frac{1}{0.165} = 6.06$ and $\frac{x}{L} = 0$ (mid – plane).

$$\frac{T_c - T_\infty}{T_o - T_\infty} = 0.6 \quad \text{[from Heisler charts]}$$

Substituting the values, we have

$$\frac{T_c - 50}{440 - 50} = 0.6$$

Or

$$T_c = 50 + 0.6(440 - 50) = 248°C$$

Temperature inside the plate 15mm from the mid-plane, $T(t) = ?$ the distance 15mm from the mid – plane implies that:

$$\frac{x}{L} = \frac{15}{30} = 0.5$$

Corresponding to $\frac{x}{L} = 0.5$ and $\frac{1}{Bi} = 6.06$, from Fig. (3.5), we have:

$$\frac{T(t) - T_\infty}{T_c - T_\infty} = 0.97$$

Substituting the values, we get:

$$\frac{T(t) - 50}{284 - 50} = 0.97$$

Or

$$T(t) = 50 + 0.97(284 - 50) = 276.98°C$$

Example (2):

A 6 mm thick stainless steel plate ($\rho = 7800 kg/m^3, c_p = 460 J/kg°C, k = 55 w/m°C$) is used to form the nose section of a missile. It is held initially at a uniform temperature of $30°C$. When the missile enters the denser layers of the atmosphere at a very high velocity the effective temperature of air surrounding the nose region attains $2150°C$; the surface convective heat transfer coefficient is estimated $3395 w/m^{2°}C$. If the maximum metal temperature is not to exceed $1100°C$, determine:

(i) Maximum permissible time in these surroundings.
(ii) Inside surface temperature under these conditions.

Solution:

Given: $2L = 6mm = 0.006m$, $k = 55w/m°C$, $c_p = 460 J/kg°$ C, $T_o = 30°C$, $\rho = 7800 kg/m^3$, $T_\infty = 2150°C$, $T(t) = 1100°C$

(i) Maximum permissible time, $\tau = ?$

Characteristic length, $L_c = \frac{0.006}{2} = 0.003m$

Biot number, $Bi = \dfrac{hL}{k} = \dfrac{3395 \times 0.003}{55} = 0.185$

As $Bi > 0.1$, therefore, lumped analysis cannot be applied in this case. Further, as $Bi < 100$, Heisler charts can be used to obtain the solution of the problem.

Corresponding to $\dfrac{1}{Bi} = 5.4$ and $\dfrac{x}{L} = 1$ (outside surface of nose section, from Fig. (3.5), we have),

$$\dfrac{T(t) - T_\infty}{T_c - T_\infty} = 0.93$$

Also,

$$\dfrac{\theta}{\theta_o} = \dfrac{T(t) - T_\infty}{T_o - T_\infty} = \left[\dfrac{T_c - T_\infty}{T_o - T_\infty}\right] \times \left[\dfrac{T(t) - T_\infty}{T_c - T_\infty}\right]$$

Or

$$\dfrac{1100 - 2150}{30 - 2150} = \left[\dfrac{T_c - T_\infty}{T_o - T_\infty}\right] \times 0.93$$

Or

$$\dfrac{T_c - T_\infty}{T_o - T_\infty} = \dfrac{1}{0.93}\left[\dfrac{1100 - 2150}{30 - 2150}\right] = 0.495$$

Now, from Fig. (3.2), corresponding to the above dimensionless temperature and $\dfrac{1}{Bi} = 5$, we got the value of Fourier number, $Fo = 4.4$

$$\therefore \dfrac{\alpha \tau}{L_c^2} = 4.4$$

Or

$$\left[\dfrac{k}{\rho c_p}\right]\left[\dfrac{\tau}{L_c^2}\right] = 4.4$$

Or

$$\left[\dfrac{55}{7800 \times 460}\right]\left[\dfrac{\tau}{0.003^2}\right] = 4.4$$

Or

$$\tau = \dfrac{4.4 \times 0.003^2 \times 7800 \times 460}{55} = 2.58s$$

(ii) Inside surface temperature, $T_c =?$

The temperature T_c at the inside surface ($x = 0$) is given by:

$$\frac{T_c - T_\infty}{T_o - T_\infty} = 0.495$$

Or

$$\frac{T_c - 2150}{30 - 2150} = 0.495$$

Or

$$T_c = 2150 + 0.495(30 - 2150) = 1100.6°C$$

Example (3):

Along cylindrical bar ($k = 17.4 w/m° C$, $\alpha = 0.019 m^2/h$) of radius $80mm$ comes out of an oven at $830° C$ throughout and is cooled by quenching it in a large bath of $40° C$ coolant. The surface coefficient of heat transferred between the bar surface and the coolant is $180 w/m^2 °C$. Determine:

(i) The time taken by the shaft center to reach $120° C$.

(ii) The surface temperature of the shaft when its center temperature is $120° C$. Also, calculate the temperature gradient at outside surface at the same instant of time.

Solution:

Given: $R = 80mm = 0.08m$, $T_o = 830°C$, $T_\infty = 40°C$, $h = 180 w/m^2 °C$, $T(t) = 120° C, k = 17.4 w/m° C, \alpha = 0.019 \, m^2/h$.

(i) The time taken by the shaft center to reach $120°C$, $\tau = ?$

Characteristic length, $L_c = \frac{\pi R^2 L}{2\pi R L} = \frac{R}{2} = \frac{0.08}{2} = 0.04m$

Biot number, $Bi = \frac{hL_c}{k} = \frac{180 \times 0.04}{17.4} = 0.413$

As $Bi > 0.1$, therefore, lumped analysis cannot be applied in this case. Further, as $Bi < 100$, Heisler charts can be used to obtain the solution of the problem.

The parametric values for the cylindrical bar are:

$$\frac{1}{Bi} = \frac{1}{0.413} = 2.42$$

$$\frac{T(t) - T_\infty}{T_o - T_\infty} = \frac{120 - 40}{830 - 40} = 0.1$$

At the center of the bar, $\frac{r}{R} = 0$

Corresponding to the above values, from the chart for an infinite cylinder Fig. (3.3), we read the Fourier number $Fo = 3.2$.

$$\therefore \frac{\alpha \tau}{(L_c)^2} = 3.2 \quad or \quad \frac{0.019 \times \tau}{0.04^2} = 3.2$$

Or

$$\tau = \frac{3.2 \times 0.04^2}{0.019} = 0.2695h \quad or \quad 970.2s$$

(ii) Temperature at the surface, $T(t)$ or $T_s = ?$

Corresponding to $\frac{r}{R} = 1$; $\frac{1}{Bi} = 2.42$, from the chart Fig. (3.6) for an infinite cylinder, we read,

$$\frac{T(t) - T_\infty}{T_c - T_\infty} = 0.83$$

Or

$$\frac{T(t) - 40}{120 - 40} = 0.83$$

Or

$$T(t) - 40 = 0.83(120 - 40)$$

Or

$$T(t) \; or \; T_s = 40 + 0.83(120 - 40) = 106.4°C$$

Temperature gradient at the outer surface, $\frac{\partial T_s}{\partial r} = ?$

$\frac{\partial T_s}{\partial r}$ at the outside surface is determined by the boundary condition $r = R$, at which, rate of energy conducted to the fluid – solid surface interface from within the solid = rate at which energy is convected away into the fluid.

$$kA_s \frac{\partial T}{\partial r} = hA_s(T_s - T_\infty)$$

Or

$$k\frac{\partial T}{\partial r} = h(T_s - T_\infty)$$

Or

$$\frac{\partial T}{\partial r} = \frac{h}{k}(T_s - T_\infty)$$

Or

$$\frac{\partial T_s}{\partial r} = \frac{180}{17.4}(106.4 - 40) = 686.89 \,°C$$

Example (4):

A $120mm$ diameter apple $\rho = 990 kg/m^3$, $c_\rho = 4170 J/kg°\,C$, $k = 0.58 w/m°C$), approximately spherical in shape is taken from a $25°C$ environment and placed in a refrigerator where temperature is $6°C$ and the average convective heat transfer coefficient over the apple surface is $12.8 w/m^{2°}C$. Determine the temperature at the center of the apple after a period of 2 hours.

Solution:

Given: $R = 120/2 = 60mm = 0.06m$, $\rho = 990 kg/m^3$, $c_\rho = 4170 J/kg°C$,
$k = 0.58 w/m°C, T_o = 25°C, T_\infty = 6°C, h = 18.8 w/m^{2°}C, \tau = 2 hours = 7200s$

The characteristic length, $L_c = \frac{\frac{4}{3}\pi R^3}{4\pi R^2} = \frac{R}{3} = \frac{0.06}{3} = 0.02m$

Biot number, $Bi = \frac{hL_c}{k} = \frac{12.8 \times 0.02}{0.58} = 0.441$,

Since $Bi > 0.1$, a lumped capacitance approach is appropriate.

Further, as $Bi < 100$, Heisler charts can be used to obtain the solution of the problem.

The parametric values for the spherical apple are:

$$\frac{1}{Bi} = \frac{1}{0.441} = 2.267$$

$$Fo = \frac{\alpha \tau}{(L_c)^2} = \left[\frac{k}{\rho c_\rho}\right]\frac{\tau}{L_c^2} = \left[\frac{0.58}{990 \times 4170}\right] \times \left[\frac{7200}{0.02^2}\right] = 0.281$$

$$\frac{r}{R} = 0 \text{ (mid} - \text{plane or center of the apple)}$$

Corresponding to the above values, from the chart for a sphere Fig. (3.7), we read

$$\frac{T_c - T_\infty}{T_o - T_\infty} = 0.75$$

Or

$$\frac{T_c - 6}{25 - 6} = 0.75$$

Or

$$T_c = 6 + 0.75(25 - 6) = 20.25 \,°C$$

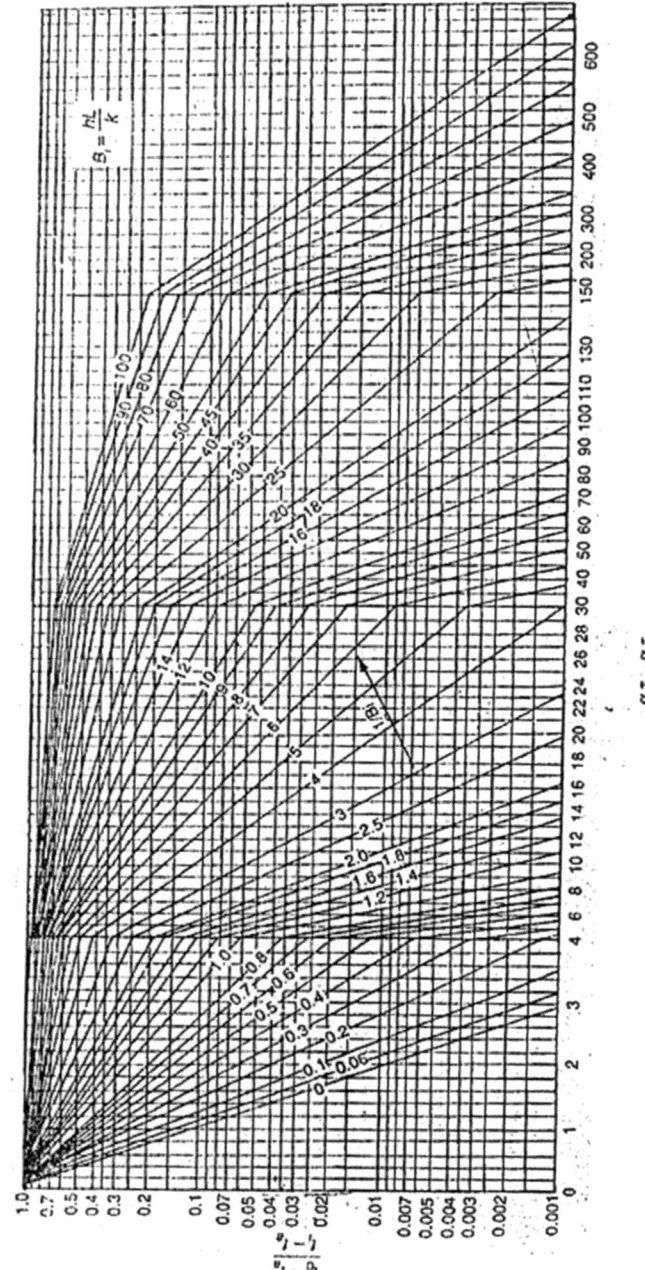

Fig. (3.2) Heisler chart for temperature history at the center of a plate of thickness 2L or (x/L) = 0

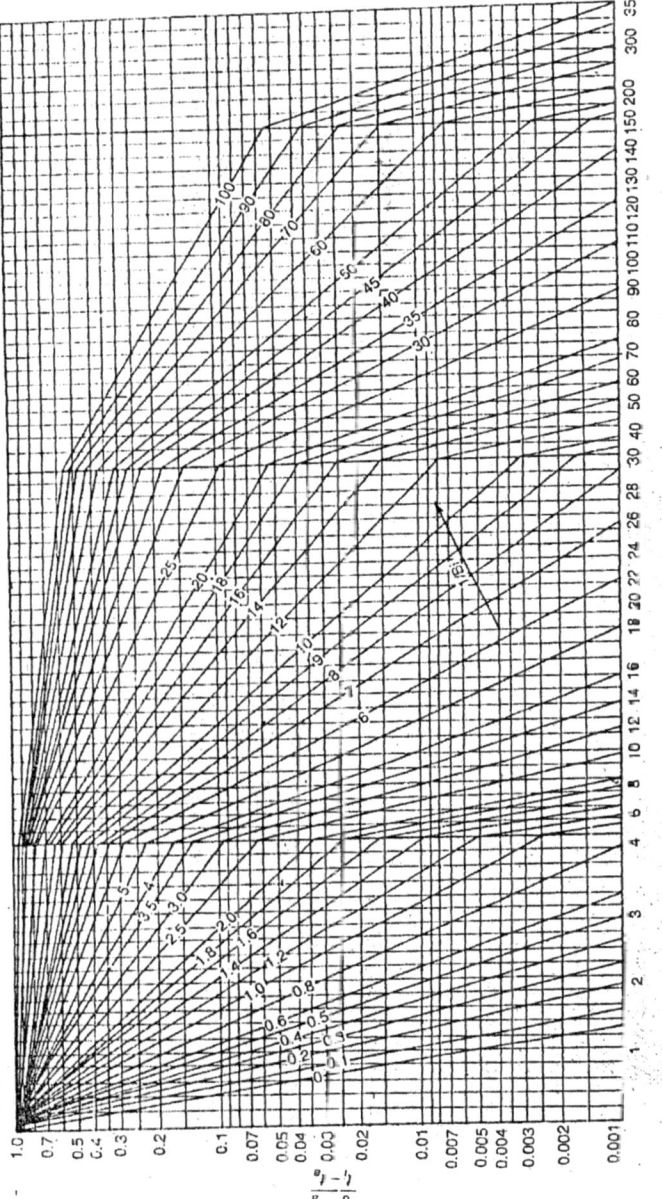

Fig. (3.3) Heisler chart for temperature history in a cylinder

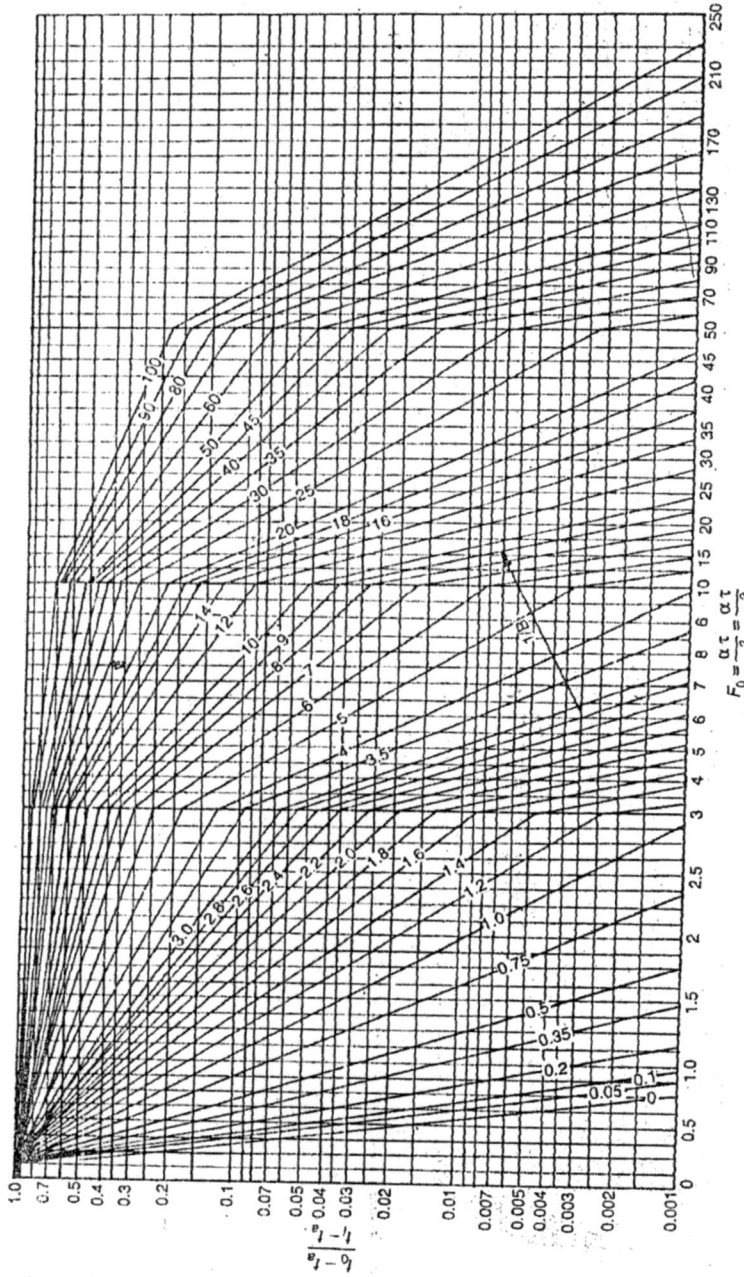

Fig. (3.4) Heisler chart for temperature history in a sphere

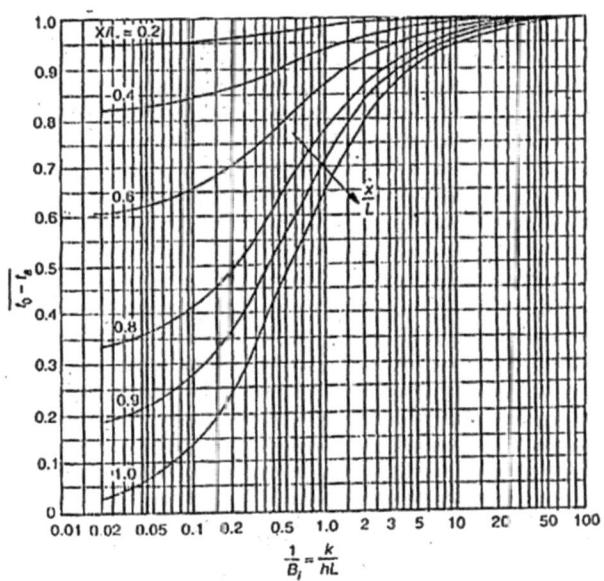

Fig. (3.5) Heisler position – correction factor chart for temperature history in plate

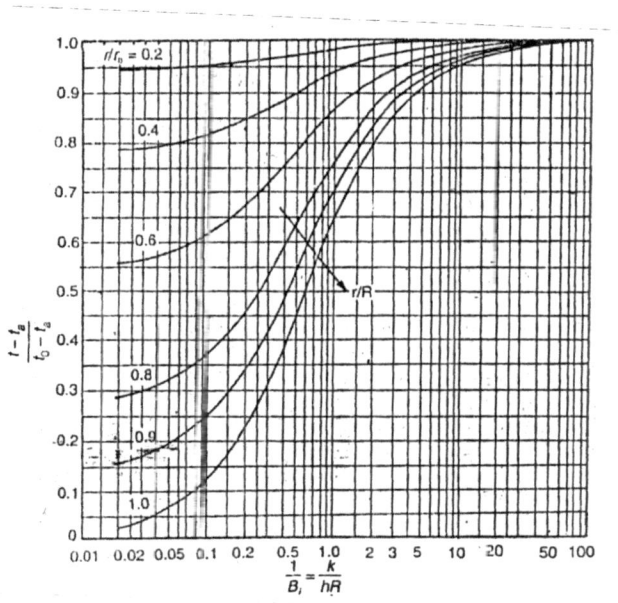

Fig. (3.6) Heisler position – correction factor chart for temperature history in cylinder

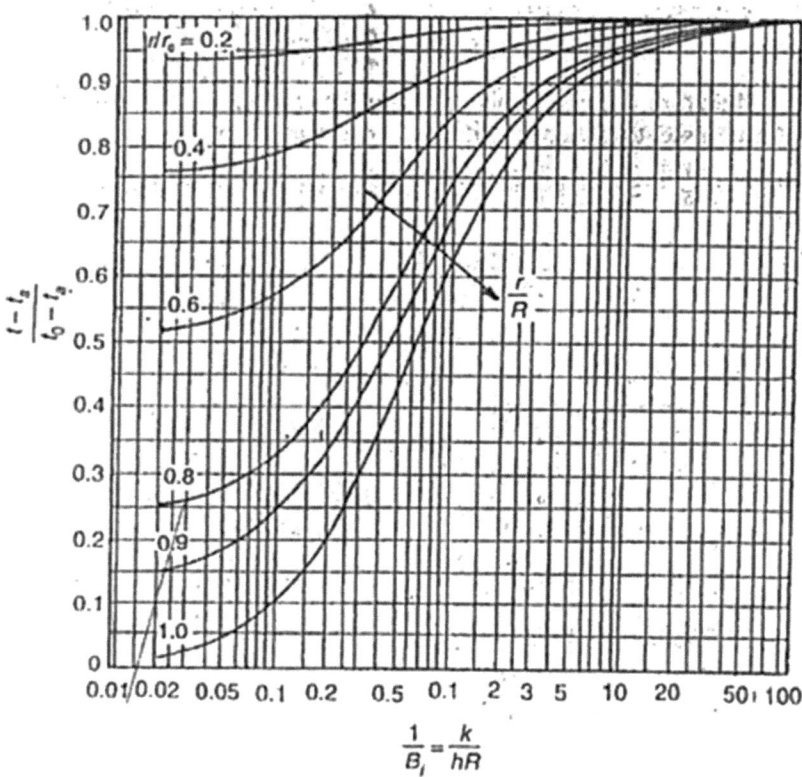

Fig. (3.7) Heisler position – correction factor chart for temperature history in cylinder

Chapter Four

Transient Heat conduction in semi – infinite solids
$[H \text{ or } Bi \to \infty]$

4.1 Introduction

A solid which extends itself infinitely in all directions of space is termed as an infinite solid. If an infinite solid is split in the middle by a plane, each half is known as semi – infinite solid. In a semi – infinite body, at any instant of time, there is always a point where the effect of heating (or cooling) at one of its boundaries is not felt at all. At the point the temperature remains unaltered. The transient temperature change in a plane of infinitely thick wall is similar to that of a semi – infinite body until enough time has passed for the surface temperature effect to penetrate through it.

As shown in Fig. (4.1) below, consider a semi – infinite plate, a plate bounded by a plane $x = 0$ and extending to infinity in the (+ve) x – direction. The entire body is initially at uniform temperature T_o including the surface at $x = 0$. The surface temperature at $x = 0$ is suddenly raised to T_∞ for all times greater than $\tau = 0$. The governing equation is:

$$\frac{d^2 t}{dx^2} = \frac{1}{\alpha} \frac{dt}{d\tau} \qquad (4.1)$$

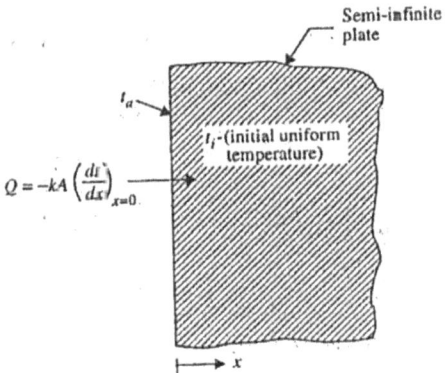

Fig. (4.1) Transition heat flow in a semi – infinite plate

The boundary conditions are:

(i) $T(x, 0) = T_o$;

(ii) $T(0, \tau) = T_\infty$ for $\tau > 0$;

(iii) $T(\infty, \tau) = T_o$ for $\tau > 0$;

The solution of the above differential equation, with these boundary conditions, for temperature distribution at any time τ at a plane parallel to and at a distance x from the surface is given by:

$$\frac{T(x, \tau) - T_\infty}{T_o - T_\infty} = \text{erf}(z) = \text{erf}\left[\frac{x}{2\sqrt{\alpha\tau}}\right] \qquad (4.2)$$

Where $z = \frac{x}{2\sqrt{\alpha\tau}}$ is known as Gaussian error function and is defined by:

$$\text{erf}\left[\frac{x}{2\sqrt{\alpha\tau}}\right] = \text{erf}(z) = \frac{2}{\sqrt{\pi}} \int_0^z e^{-\eta^2} d\eta \qquad (4.3)$$

With $\text{erf}(0) = 0, \text{erf}(\infty) = 1$.

Table (4.1) shows a few representative values of erf(z). Suitable values of error functions may be obtained from Fig. (4.2) below.

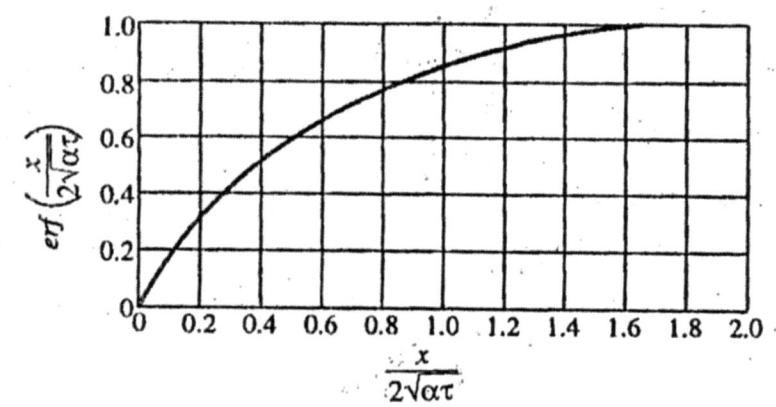

Fig. (4.2) Gauss's error integral

Table (4.1) The error function

$$\text{erf}(z) = \frac{2}{\sqrt{\pi}} \int_c^z e^{-\eta^2} d\eta \quad \text{where } z = \frac{x}{2\sqrt{\alpha\tau}}$$

z	erf(z)	z	erf(z)
0.00	0.0000	0.32	0.3491
0.02	0.0225	0.34	0.3694
0.04	0.0451	0.36	0.3893
0.06	0.0676	0.38	0.4090
0.08	0.0901	0.40	0.4284
0.10	0.1125	0.42	0.4475
0.12	0.1348	0.44	0.4662
0.14	0.1569	0.46	0.4847
0.16	0.1709	0.48	0.5027
0.18	0.2009	0.50	0.5205
0.20	0.2227	0.55	0.5633
0.22	0.2443	0.60	0.6039
0.024	0.2657	0.65	0.6420
0.26	0.2869	0.70	0.6778
0.28	0.3079	0.75	0.7112
0.30	0.3286	0.80	0.7421
0.85	0.7707	1.65	0.9800
0.90	0.7970	1.70	0.9883
0.95	0.8270	1.75	0.9864
1.00	0.8427	1.80	0.9891
1.05	0.8614	1.85	0.9909
1.10	0.8802	1.90	0.9928
1.15	0.8952	1.95	0.9940
1.20	0.9103	2.00	0.9953
1.25	0.9221	2.10	0.9967
1.30	0.9340	2.20	0.9981
1.35	0.9431	2.30	0.9987
1.40	0.9523	2.40	0.9993
1.45	0.9592	2.50	0.9995
1.50	0.9661	2.60	0.9998
1.55	0.9712	2.80	0.9999
1.60	0.9753	3.00	1.0000

By insertion of definition of error function in equation (4.2), we get

$$T(x,\tau) = T_\infty + (T_o - T_\infty)\frac{2}{\sqrt{\pi}} \int_0^z e^{-\eta^2} d\eta$$

On differentiating the above equation, we obtain

$$\frac{\partial T}{\partial x} = \frac{T_o - T_\infty}{\sqrt{\pi \alpha \tau}} e^{[-x^2/(4\alpha\tau)]}$$

∴ The instantaneous heat flow rate at a given x – location within the semi – infinite body at a specified time is given by:

$$Q_{instantaneous} = -kA(T_o - T_\infty)\frac{e^{\left[-\frac{x^2}{4\alpha\tau}\right]}}{\sqrt{\pi\alpha\tau}} \quad (4.4)$$

By substituting the gradient $\left[\frac{\partial T}{\partial x}\right]$ in Fourier's law.

The heat flow rate at the surface ($x = 0$) is given by:

$$Q_{surface} = \frac{-kA(T_o - T_\infty)}{\sqrt{\pi\alpha\tau}} \quad (4.5)$$

∴ The total heat flow rate,

$$Q(t) = \frac{-kA(T_o - T_\infty)}{\sqrt{\pi\alpha}}\int_0^\tau \frac{1}{\sqrt{\tau}}d\tau = -kA(T_o - T_\infty)2\sqrt{\frac{\tau}{\pi\alpha}}$$

Or

$$Q(t) = -1.13 kA(T_o - T_\infty)\sqrt{\frac{\tau}{\alpha}} \quad (4.6)$$

The general criterion for the infinite solution to apply to a body of finite thickness (slab) subjected to one dimensional heat transfer is:

$$\frac{L}{2\sqrt{\alpha\tau}} \geq 0.5$$

Where, L = thickness of the body.

The temperature at the center of cylinder or sphere of radius R, under similar conditions of heating or cooling, is given as follows:

$$\frac{T(t) - T_\infty}{T_o - T_\infty} = erf\left[\frac{\alpha\tau}{R^2}\right] \quad (4.7)$$

For the cylindrical and spherical surfaces the values of function $erf\left[\frac{\alpha\tau}{R^2}\right]$ can be obtained from Fig. (4.3) which is shown below.

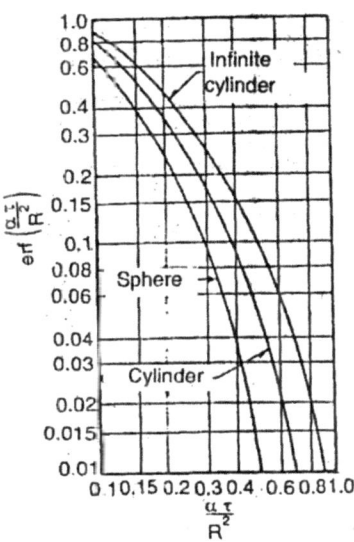

Fig. (4.3) Error integral for cylinders and spheres

4.2 Penetration Depth and Penetration Time

Penetration depth refers to the location of a point where the temperature change is within 1 percent of the change in the surface temperature.

$$i.e. \quad \frac{T(t) - T_\infty}{T_o - T_\infty} = 0.9$$

This corresponds to $\frac{x}{2\sqrt{\alpha\tau}} = 1.8$, from the table for Gaussian error integral.

Thus, the depth (d) to which the temperature perturbation at the surface has penetrated,

$$d = 3.6 \sqrt{\alpha\tau}$$

Penetration time is the time τ_P taken for a surface penetration to be felt at that depth in the range of 1 percent. It is given by:

$$\frac{d}{2\sqrt{\alpha\tau_P}} = 1.8$$

Or
$$\tau_P = \frac{d^2}{13\,\alpha} \qquad (4.8)$$

4.3 Solved Examples

Example (1):

A steel ingot (large in size) heated uniformly to $745°\,C$ is hardened by quenching it in an oil path maintained at $20°\,C$. Determine the length of time required for the temperature to reach $595°\,C$ at a depth of $12mm$. The ingot may be approximated as a flat plate. For steel ingot take α(thermal diffusivity) $= 1.2 \times 10^{-5} m^2/s$

Solution:

Given: $T_o = 745°c, T_\infty = 20°c, T(t) = 595°c, x = 12mm = 0.012m,$
$\alpha = 1.2 \times 10^{-5} m^2/s,$ time required, $\tau =?$

The temperature distribution at any time τ at a plane parallel to and at a distance x from the surface is given by:

$$\frac{T(t) - T_\infty}{T_c - T_\infty} = erf\left[\frac{x}{2\sqrt{\alpha\,\tau}}\right] \qquad (4.2)$$

Or
$$\frac{595 - 20}{745 - 20} = 0.79 = erf\left[\frac{x}{2\sqrt{\alpha\,\tau}}\right]$$

Or
$$\therefore \frac{x}{2\sqrt{\alpha\,\tau}} = 0.9 \qquad \text{from Table (4.1) or Fig. (4.3)}$$

Or
$$\frac{x^2}{4\,\alpha\,\tau} = 0.81$$

Or
$$\tau = \frac{x^2}{4\,\alpha \times 0.81} = \frac{0.012^2}{4 \times 1.2 \times 10^{-5} \times 0.81} = 3.7s$$

Example (2):

It is proposed to bury water pipes underground in wet soil which is initially at $5.4°\,C$. The temperature of the surface of soil suddenly drops to $-6°\,C$ and remains at this value for 9.5 hours. Determine the maximum depth at which the pipes be laid if the surrounding soil temperature is to remain above $0°\,C$ (without water getting frozen). Assume the soil as semi – infinite solid.

For wet soil take α(thermal diffusivity) $= 2.75 \times 10^{-3} m^2/h$

Solution:

Given: $T_o = 5.4°C, T_\infty = -6°c, T_{(t)} = 0°C, \alpha = 2.75 \times 10^{-3} m^2/s$, maximum depth $x = ?$

The temperature, at critical depth, will just reach after 9.5 hours,

Now,

$$\frac{T(t) - T_\infty}{T_c - T_\infty} = erf\left[\frac{x}{2\sqrt{\alpha\,\tau}}\right] \quad (4.2)$$

Or

$$\frac{0 - (-6)}{5.4 - (-6)} = 0.526 = erf\left[\frac{x}{\sqrt{2\,\alpha\,\tau}}\right]$$

Or

$$\frac{x}{2\sqrt{\alpha\,\tau}} \simeq 0.5 \quad \text{from Table (4.1) or Fig. (4.3)}$$

Or

$$x = 0.5 \times 2\sqrt{\alpha\,\tau}$$

Or

$$x = 0.5 \times 2\sqrt{2.75 \times 10^{-3} \times 9.5} = 0.162 m$$

Example (3):

A $60mm$ thick mild steel plate ($\alpha = 1.22 \times 10^{-5} m^2/s$) is initially at a temperature of $30°\,C$. It is suddenly exposed on one side to a fluid which causes the surface temperature to increase to and remain at $110°\,C$. Determine:

(i) The maximum time that the slab be treated as a semi – infinite body;

(ii) The temperature at the center of the slab 1.5 minutes after the change in surface temperature.

Solution:

Given: $L = 60mm = 0.06m$, $\alpha = 1.22 \times 10^{-5} m^2/s$, $T_o = 30°C$, $T_\infty = 110°C$,

$\tau = 1.5 \, minutes = 90 \, s$

(i) The maximum time that the slab be treated as a semi – infinite body, ($\tau_{max} =$). The general criterion for the infinite solution to apply to a body of finite thickness subjected to one – dimensional heat transfer is:

$$\frac{L}{2\sqrt{\alpha \tau}} \geq 0.5 \quad \text{(where L = thickness of the body)}$$

Or

$$\frac{L}{2\sqrt{\alpha \tau_{max}}} = 0.5 \quad \text{or} \quad \frac{L^2}{4\alpha \tau_{max}} = 0.25$$

Or

$$\tau_{max} = \frac{L^2}{4\alpha \times 0.25} = \frac{0.06^2}{4 \times 1.22 \times 10^{-5} \times 0.25} = 295.1 \, s$$

(ii) The temperature at the center of the slab, $T(t) = ?$

At the center of the slab, $x = 0.03m$; $\tau = 90 \, s$

$$\frac{T(t) - T_\infty}{T_o - T_\infty} = erf\left[\frac{x}{2\sqrt{\alpha \tau}}\right]$$

Or

$$T(t) = T_\infty + erf\left[\frac{x}{2\sqrt{\alpha \tau}}\right](T_o - T_\infty)$$

Where:

$$erf\left[\frac{x}{2\sqrt{\alpha \tau}}\right] = erf\left[\frac{0.03}{2\sqrt{1.22 \times 10^{-5} \times 90}}\right] = erf(0.453)$$

$\simeq 0.47$ [from Table (4.1)]

$$T(t) = T_c = 110 + 0.47(30 - 110) = 72.4°C$$

Example (4):

The initial uniform temperature of a thick concrete wall ($\alpha = 1.6 \times 10^{-3} m^2/h$, $k = 0.9 w/m°C$) of a jet engine test cell is $25°C$. The surface temperature of the wall suddenly rises to $340°C$ when the combination of exhaust gases from the turbo jet … spray of cooling water occurs. Determine:

(i) The temperature at a point 80mm from the surface after 8 hours.

(ii) The instantaneous heat flow rate at the specified plane and at the surface itself at the instant mentioned at (i).

Use the solution for semi – infinite solid.

Solution:

Given: $T_o = 25°C$, $T_\infty = 340°C$, $\alpha = 1.6 \times 10^{-3} m^2/h$, $k = 0.94 w/m°C$, $\tau = 8h$,

$x = 80mm = 0.08m$

(i) The temperature at a point 0.08m from the surface; $T(t) = ?$

$$\frac{T(t) - T_\infty}{T_o - T_\infty} = erf\left[\frac{x}{2\sqrt{\alpha \tau}}\right]$$

Or

$$T(t) = T_\infty + erf\left[\frac{x}{2\sqrt{\alpha \tau}}\right](T_o - T_\infty)$$

Where

$$erf\left[\frac{x}{2\sqrt{\alpha \tau}}\right] = erf\left[\frac{0.03}{2\sqrt{1.6 \times 10^{-3} \times 8}}\right] = erf(0.353) \simeq 0.37$$

$$\therefore T(t) = 340 + 0.37(25 - 340) = 223.45°C$$

(ii) The instantaneous heat flow rate, $Q_{instantaneous}$ at the specified plane =?

$$Q_i = -kA(T_o - T_\infty)\frac{e^{\left[-\frac{x^2}{4\alpha\tau}\right]}}{\sqrt{\pi \alpha \tau}} \quad \text{from equation (4.4)}$$

$$Q_i = -0.94 \times 1 \times (25 - 340)\frac{e^{[-0.08^2/(4\times 1.6\times 10^{-3}\times 8)]}}{\sqrt{\pi \times 1.6 \times 10^{-3} \times 8}}$$

$$= -296.1 \times \frac{0.8825}{0.2005} = -1303.28 w/m^2 \text{ of wall area}$$

The negative sign shows the heat lost from the wall.
Heat flow rate at the surface itself, $Q_{surface} = ?$

$$Q_{surface} \text{ or } Q_s = -\frac{kA(T_o - T_\infty)}{\sqrt{\pi \alpha \tau}} \quad \text{from equation(4.5)}$$

$$= -\frac{0.94 \times 1 \times (25 - 340)}{\sqrt{\pi \times 1.6 \times 10^{-3} \times 8}} = (-)1476.6w \text{ per } m^2 \text{ of wall area}$$

Example (5):

The initial uniform temperature of a large mass of material ($\alpha = 0.42 m^2/h$) is $120°C$. The surface is suddenly exposed to and held permanently at $6°C$. Calculate the time required for the temperature gradient at the surface to reach $400°C/m$.

Solution:

Given: $T_o = 120°C, T_\infty = 6°C, \alpha = 0.42 m^2/h,$

$\left[\frac{\partial T}{\partial x}\right]_{x=0}$ (temperature gradient at the surface) $= 400°C/m$

Time required, $\tau = ?$

Heat flow rate at the surface ($x = 0$) is given by:

$$Q_{surface} = -\frac{kA(T_o - T_\infty)}{\sqrt{\pi \alpha \tau}} \quad \text{from equation (4.5)}$$

Or

$$-kA\left[\frac{\partial T}{\partial x}\right]_{x=0} = -\frac{kA(T_o - T_\infty)}{\sqrt{\pi \alpha \tau}}$$

Or

$$\left[\frac{\partial T}{\partial x}\right]_{x=0} = \frac{T_o - T_\infty}{\sqrt{\pi \alpha \tau}}$$

substituting the values above, we obtain:

$$400 = \frac{(120 - 6)}{\sqrt{\pi \times 0.42 \times \tau}}$$

Or

$$\pi \times 0.42\tau = \left[\frac{120-6}{400}\right]^2 = 0.0812$$

Or

$$\tau = \frac{0.0812}{\pi \times 0.42} = 0.0615h = 221.4\ s$$

Example (6):

A motor car of mass 1600kg travelling at 90km/h, is brought to reset within a period of 9 seconds when the brakes are applied. The braking system consists of 4 brakes with each brake band of 360 cm² area, these press against steel drums of equivalent area. The brake lining and the drum surface ($k = 54w/m°C, \alpha = 1.25 \times 10^{-5} m^2/s$) are at the same temperature and the heat generated during the stoppage action dissipates by flowing across drums. The drum surface is treated as semi – infinite plane, calculate the maximum temperature rise.

Solution:

Given: $m = 1600kg, v(\text{velocity}) = 90km/h, \tau = 9s, A(\text{Area of 4 brake bands})$
$= 4 \times 360 \times 10^{-4} m^2$ or $0.144 m^2, k = 54w/m°C, \alpha = 1.25 \times 10^{-5} m^2/s$.

Maximum temperature rise, $T_\infty - T_o =?$

When the car comes to rest (after applying brakes), its kinetic energy is converted into heat energy which is dissipated through the drums.

Kinetic energy of the moving car $= \frac{1}{2}mv^2$

$$= \frac{1}{2} \times 1600 \times \left[\frac{90 \times 1000}{60 \times 60}\right]^2$$

$$= 5 \times 10^5 J \text{ in 9 seconds}$$

$$\therefore \text{Heat flow rate} = \frac{5 \times 10^5}{9} = 0.555 \times 10^5 J/s\ or\ w$$

This value equals the instantaneous heat flow rate at the surface ($x = 0$), which is given by:

$$(Q_i)_{surface} = -\frac{kA(T_o - T_\infty)}{\sqrt{\pi \alpha \tau}} = 0.555 \times 10^5 \quad \text{from equation (4.5)}$$

Or

$$-\frac{54 \times 0.144(T_o - T_\infty)}{\sqrt{\pi \times 1.25 \times 10^{-5} \times 9}} = 0.555 \times 10^5$$

Or

$$-(T_o - T_\infty) = \frac{0.555 \times 10^5 \times \sqrt{\pi \times 1.25 \times 10^{-5} \times 9}}{54 \times 0.144} = 134.3$$

Or

$$T_o - T_\infty = 134.3°C$$

Hence, maximum temperature rise = $134.3°C$

Example (7):

A copper cylinder ($\alpha = 1.12 \times 10^{-4} m^2/s$), $600mm$ in diameter and $750mm$ in length, is initially at a uniform temperature of $20°C$. When the cylinder is exposed to hot flue gases, its surface temperature suddenly increases to $480°C$. Calculate:

(i) The temperature at the center of cylinder 3 minutes after the operation of change in surface temperature;

(ii) Time required to attain a temperature of $350°C$.

Assume the cylinder as semi – infinite solid.

Solution:

Given: $R = \frac{600}{2} = 300mm$ or $0.3m$, $\alpha = 1.12 \times 10^{-4} m^2/s$, $T_o = 20°C$,
$T_\infty = 480°C$, $T(t) = 350°C$, $\tau = 3 \times 60 = 180\ s$

(i) The temperature at the center of the cylinder, $T(t)$ or $T_c = ?$

The temperature distribution at the center of the cylinder is expressed as:

$$\frac{T(t) - T_\infty}{T_c - T_\infty} = erf\left[\frac{\alpha \tau}{R^2}\right] \quad \text{from equation (4.7)}$$

Where:

$$erf\left[\frac{\alpha\tau}{R^2}\right] = erf\left[\frac{1.12\times 10^{-4}\times 180}{0.3^2}\right] = erf(0.224) \simeq 0.32 \text{ [from Fig. (4.3)]}$$

Substituting the values, we obtain:

$$\frac{T(t)-480}{20-480} = 0.32$$

Or

$$T(t) = 480 + 0.32(20-480) = 332.8°C$$

(ii) Time required to attain a temperature of $350°C$, $\tau =?$

$$\frac{350-480}{20-480} = erf\left[\frac{\alpha\tau}{R^2}\right]$$

$$0.2826 = erf\left[\frac{\alpha\tau}{R^2}\right]$$

$$\therefore \frac{\alpha\tau}{R^2} \simeq 0.23$$

Or

$$\tau = \frac{0.23\times R^2}{\alpha} = \frac{0.23\times 0.3^2}{1.12\times 10^{-4}} = 184.8\ s$$

Chapter Five

Systems with Periodic Variation of Surface Temperature

5.1 Introduction

The periodic type of heat flow occurs in cyclic generators, in reciprocating internal combustion engines and in the earth as the result of daily cycle of the sun. These periodic changes, in general, are not simply sinusoidal but rather complex. However, these complex changes can be approximated by a number of sinusoidal components.

Let us consider a thick plane wall (one dimensional case) whose surface temperature alters according to a sine function as shown in Fig. (5.1) below. The surface temperature oscillates about the mean temperature level t_m according to the following relation:

$$\theta_{s,\tau} = \theta_{s,a} \sin(2\pi n\tau)$$

Where,

$\theta_{s,\tau}$ = excess over the mean temperature $(= t_{s,\tau} - t_m)$;

$\theta_{s,a}$ = Amplitude of temperature excess, i.e., the maximum temperature excess at the surface;

n = Frequency of temperature wave.

The temperature excess at any depth x and time τ can be expressed by the following relation:

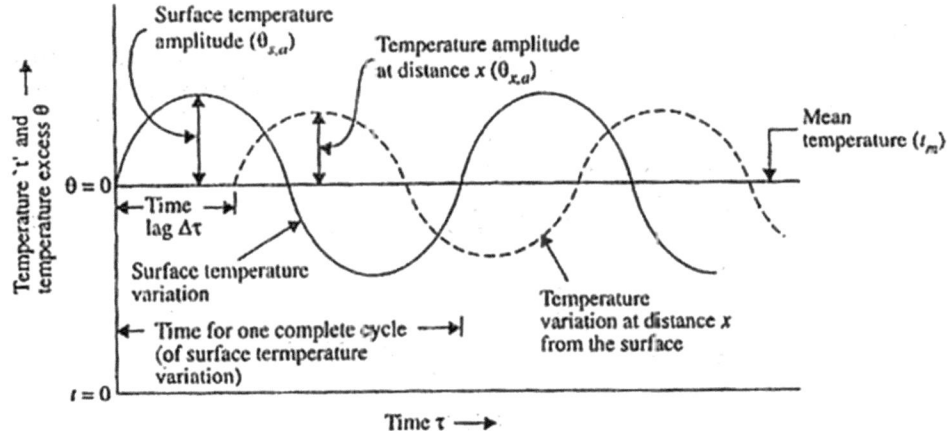

Fig. (5.1) Temperature curves for periodic variation of surface temperature

$$\theta_{x,\tau} = \theta_{s,a} \exp\left[-x\sqrt{\pi n/\alpha}\right] \sin\left[2\pi n\tau - x\sqrt{\frac{\pi n}{\alpha}}\right] \quad (5.1)$$

The temperature excess, at the surface ($x = 0$), becomes zero at $\tau = 0$. But at any depth, $x > 0$, a time $\left[\left[\frac{x}{2}\right]\left[\frac{1}{\sqrt{\alpha\pi n}}\right]\right]$ would elapse before the temperature excess $\theta_{x,\tau}$ becomes zero. The time interval between the two instant is called the time lag.

$$\text{The time lag } \Delta\tau = \frac{x}{2}\sqrt{\frac{1}{\alpha\pi n}} \quad (5.2)$$

At depth x, the temperature amplitude ($\theta_{x,a}$) is given by:

$$\theta_{x,a} = \theta_{s,c} \exp\left[-x\sqrt{\frac{\pi n}{\alpha}}\right] \quad (5.3)$$

The above relations indicate the following facts:

1. At any depth, $x > 0$, the amplitude (maximum value) occurs late and is smaller than that at the surface ($x = 0$).

2. The amplitude of temperature oscillation decreases with increasing depth. (Therefore, the amplitude becomes negligibly small at a particular depth inside

the solid and consequently a solid thicker than this particular depth is not of any importance as far as variation in temperature is concerned).
3. With increasing value of frequency, time lag and the amplitude reduce.
4. Increase in diffusivity α decreases the time lag but keeps the amplitude large.
5. The amplitude of temperature depends upon depth x as well as the factor $\sqrt{\frac{n}{\alpha}}$. Thus, if $\sqrt{\frac{n}{\alpha}}$ is large, equation (5.3) holds good for thin solid rods also.

5.2 Solved Examples

Example (1):

During the periodic heating and cooling of a thick brick wall, the wall temperature varies sinusoidally. The surface temperature ranges from $30°C$ to $80°C$ during a period of 24 hours. Determine the time lag of the temperature wave corresponding to a point located at 300mm from the wall surface.

The properties of the wall material are:

$\rho = 1610 kg/m^3, k = 0.65 w/m°C; c_P = 440 J/kg°C$

Solution:

Given: $x = 300mm = 0.3m, \rho = 1610 kg/m^3, k = 0.65 w/m°C;$

$c_P = 440 J/kg°C, n = \frac{1}{24} = 0.04167/h$

Time lag $\Delta \tau = ?$

$$\Delta \tau = \frac{x}{2}\sqrt{\frac{1}{\alpha \pi n}} \qquad \text{from equation (5.2)}$$

Where:

$$\alpha = \frac{k}{\rho c_P} = \frac{0.65}{1610 \times 440} = 9.176 \times 10^{-7} \, m^2/s \quad \text{or} \quad 0.0033 m^2/h$$

$$\therefore \Delta \tau = \frac{0.3}{2} \times \sqrt{\frac{1}{0.0033 \times \pi \times 0.04167}} = 7.2 \, h$$

Example (2):

A single cylinder ($\alpha = 0.044 m^2/h$ for cylinder material) two – stroke I.C. engine operates at 1400 rev/min. Calculate the depth where the temperature wave due to variation of cylinder temperature is damped to 2% of its surface value.

Solution:

Given: $\alpha = 0.044 m^2/h$, $n = 1400 \times 60 = 84000/h$

The amplitude of temperature excess, at any depth x, is given by:

$$\theta_{x,a} = \theta_{s,a}\, e^{-x\sqrt{\frac{\pi n}{\alpha}}} \qquad \text{from equation (5.3)}$$

Or

$$\frac{\theta_{x,a}}{\theta_{s,a}} = e^{-x\sqrt{\frac{\pi n}{\alpha}}}$$

Or

$$\frac{2}{100} = e^{-x\sqrt{\frac{\pi \times 84000}{0.044}}} = e^{-2449 x}$$

$$\ln 0.02 = -2449 \times \ln e$$

$$x = \frac{\ln 0.02}{-2449 \times 1} = 1.597 \times 10^{-3} m \quad \text{or} \quad 1.597 mm$$

Chapter Six

Transient Conduction with Given Temperature Distribution

6.1 Introduction

The temperature distribution at some instant of time, in some situations, is known for the one – dimensional transient heat conduction through a solid.

The known temperature distribution may be expressed in the form of polynomial

$t = a - bx + cx^2 + dx^3 - ex^4$ Where a, b, c, d and e are the known coefficients.

By using such distribution, the one – dimensional transient heat conduction problem can be solved.

6.2 Solved Examples

Example (1):

The temperature distribution across a large concrete slab $500mm$ thick heated from one side as measured by thermocouples approximates to the following relation,

$t = 120 - 100x + 24x^2 + 40x^3 - 30x^4$

Where t is in $°C$ and x is in m. Considering an area of $4m^2$, Calculate:

(i) The heat entering and leaving the slab in unit time;

(ii) The heat energy stored in unit time;

(iii) The rate of temperature change at both sides of the slab;

(iv) The point where the rate of heating or cooling is maximum.

The properties of concrete are as follows:

$k = 1.2 \, w/m°C,$ $\alpha = 1.77 \times 10^{-3} m^2/h$

Solution:

Given: $A = 4m^2, x = 500mm = 0.5m, k = 1.22 w/m°C, \alpha = 1.77 \times 10^{-3} m^2/h$

$t = 120 - 100x + 24x^2 + 40x^3 - 30x^4$ (Temperature distribution polynomial)

$\dfrac{dt}{dx} = -100 + 48x + 120x^2 - 120x^3$

$$\frac{d^2t}{dx^2} = 48 + 240x - 360x^2$$

(i) The heat entering and leaving the slab in unit time: $Q_i = ?$ $Q_o = ?$

Heat leaving the slab,

$$Q_{in} = -kA \left[\frac{dt}{dx}\right]_{x=0} = (-1.2 \times 5)(-100) = 600w$$

Heat leaving the slab,

$$Q_{out} = -kA \left[\frac{dt}{dx}\right]_{x=0.5} = (-1.2 \times 5)(-100 + 48 \times 0.5 + 120 \times 0.5^2 - 120 \times 0.5^2)$$

$$= 0.6(-100 + 24 + 30 - 15) = 366w$$

(ii) The heat energy stored in unit time:

$$\text{rate of heat storage} = Q_{in} - Q_{out} = 600 - 366 = 234w$$

(iii) The rate of temperature change at both sides of the slab: $\left[\frac{dt}{d\tau}\right]_{x=0}$ and $\left[\frac{dt}{d\tau}\right]_{x=0.5} = ?$

$$\frac{dt}{d\tau} = \alpha \frac{d^2t}{dx^2} = \alpha(48 + 240x - 360x^2)$$

$$\therefore \left[\frac{dt}{d\tau}\right]_{x=0} = 1.77 \times 10^{-3}(48) = 0.08496\ °C/h$$

and, $\left[\frac{dt}{d\tau}\right]_{x=0.5} = 1.77 \times 10^{-3}(48 + 240 \times 0.5 - 360 \times 0.5^2) = 1.3806\ °c/h$

(iv) The point where the rate of heating or cooling is maximum, x:

$$\frac{d}{dx}\left[\frac{dt}{d\tau}\right] = 0$$

Or

$$\frac{d}{dx}\left[\alpha \frac{d^2t}{dx^2}\right] = 0$$

Or

$$\frac{d^3t}{dx^3} = 0$$

Or

$$240 - 720x = 0$$

$$\therefore x = \frac{240}{720} = 0.333m$$

Chapter Seven

Additional Solved Examples in Lumped Capacitance System

7.1 Example (1): Determination of Temperature and Rate of Cooling of a Steel Ball

A steel ball 100mm in diameter and initially at $900°C$ is placed in air at $30°C$, find:

(i) Temperature of the ball after 30 seconds.

(ii) The rate of cooling $(°C/min.)$ after 30 seconds.

Take: $h = 20 w/m^{2}°C$; $k(\text{steel}) = 40 w/m°C$; $\rho(\text{steel}) = 7800 kg/m^3$;
$c_P(\text{steel}) = 460 J/kg°C$

Solution:

Given: $R = \frac{100}{2} = 50mm$ or $0.05m$. $T_o = 900°C$; $T_\infty = 30°C, h = 20 w/m^{2}°C$;
$k(\text{steel}) = 40 w/m°C$; $\rho(\text{steel}) = 7800 kg/m^3$; $c_P(\text{steel}) = 460 J/kg°C$;

$\tau = 30s$

(i) Temperature of the ball after 30 seconds: $T(t) = ?$

Characteristic length,

$$L_c = \frac{V}{A_s} = \frac{\frac{4}{3}\pi R^3}{4\pi R^2} = \frac{R}{3} = \frac{0.05}{3} = 0.01667 m$$

Biot number,

$$Bi = \frac{hL_c}{k} = \frac{20 \times 0.01667}{40} = 0.008335$$

Since, Bi is less than 0.1, hence lumped capacitance method (Newtonian heating or cooling) may be applied for the solution of the problem.

The time versus temperature distribution is given by equation (1.4):

$$\frac{\theta}{\theta_o} = \frac{T(t) - T_\infty}{T_o - T_\infty} = e^{\frac{-hA_s}{\rho V c_P}} \qquad (1)$$

Now,

$$\frac{hA_s}{\rho V c_p} \cdot \tau \left[\frac{h}{\rho c_p}\right]\left[\frac{A_s}{V}\right]\tau = \left[\frac{20}{7800 \times 460}\right]\left[\frac{1}{0.01667}\right](30) = 0.01$$

$$\therefore \frac{T(t) - 30}{900 - 30} = e^{-0.01} = 0.99$$

Or

$$T(t) = 30 + 0.99(900 - 30) = 891.3°C$$

(ii) The rate of cooling (°C/min) after 30 seconds: $\frac{dt}{d\tau} = ?$

The rate of cooling means we have to find out $\frac{dt}{d\tau}$ at the required time. Now, differentiating equation (1), we get:

$$\frac{1}{T_o - T_\infty} \times \frac{dt}{d\tau} = -\left[\frac{hA_s}{\rho v c_p}\right] e^{\frac{-hA_s}{\rho V c_p}\tau}$$

Now, substituting the proper values in the above equation, we have:

$$\frac{1}{(900 - 30)} \cdot \frac{dt}{d\tau} = -\left[\frac{20}{7800 \times 460}\right] \times \frac{1}{0.01667} \times 0.99 = -3.31 \times 10^{-4}$$

$$\therefore \frac{dt}{d\tau} = (900 - 30)(-3.31 \times 10^{-4}) = -0.288°C/s$$

or $\quad \frac{dt}{d\tau} = -0.288 \times 60 = -17.28°c/mim$

7.2 Example (2): Calculation of the Time Required to Cool a Thin Copper Plate

A thin copper plate $20mm$ thick is initially at $150°C$. One surface is in contact with water at $30°c$ ($h_w = 100w/m^{2°}C$) and the other surface is exposed to air at $30°C$ ($h_a = 20w/m^{2°}C$). Determine the time required to cool the plate to $90°C$.

Take the following properties of the copper:

$\rho = 8800 kg/m^3$; $c_p = 400 J/kg°C$ and $k = 360 w/m°C$

The plate is shown in Fig. (7.1) below:

Solution:

Given: $L = 20mm$ or $0.02m$; $T_c = 150°C$; $T_\infty = 30°C$; $h_w = 100 w/m^{2°}C$; $h_a = 20w/m^{2°}C$; $T(t) = 90°C$; $\rho = 8800 kg/m^3$; $c_P = 400 J/kg°C$; $k = 360 w/m°C$

Time required to cool the plate, $\tau = ?$

Biot number,

$$Bi = \frac{hL_c}{k} = \frac{h\left(\frac{L}{2}\right)}{k} = \frac{100 \times (0.02/2)}{360} = 0.00277$$

Since, $Bi < 0.1$, the internal resistance can be neglected and lumped capacitance method may be applied for the solution of the problem.

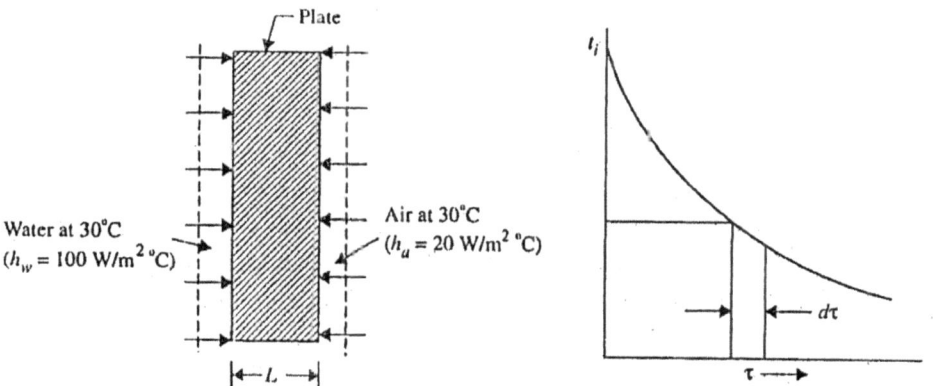

Fig. (7.1)

The basic heat transfer equation can be written as:

$$d\theta = -mc_P \frac{dt}{d\tau} = h_w A_s(T(t) - T_w) + h_a A_s(T(t) - T_a)$$

$$= A_s[h_w(T(t) - T_w) + h_a(T(t) - T_a)]$$

Where T_w and T_a are temperatures of water and air respectively and they are not changing with time.

$$\therefore -\rho A_s L\, c_P \left(\frac{dT(t)}{d\tau}\right) = A_s[h_w(T(t) - T_w) + h_a(T(t) - T_a)]$$

Or
$$-\rho L c_P \frac{dt}{d\tau} = T(t)(h_w + h_a) - (h_w T_w + h_a T_a)$$

Or
$$\frac{dT(t)}{T(t)(h_w + h_a) - (h_w T_w + h_a T_a)} = \frac{d\tau}{\rho L c_P}$$

Or
$$\frac{dT(t)}{c_1 T(t) - c_2} = -\frac{d\tau}{\rho L c_P}$$

Where
$$c_1 = h_w + h_a \quad \text{and} \quad c_2 = h_w T_w + h_a T_a$$

$$\therefore \frac{1}{c_1} \int \frac{dT(t)}{T(t) - \frac{c_2}{c_1}} = -\int \frac{d\tau}{\rho L c_P}$$

Or
$$\frac{1}{c_1} \int_{T_o}^{T(t)} \frac{dT(t)}{T(t) - \frac{c_2}{c_1}} = -\int_0^{\tau} \frac{d\tau}{\rho L c_P} \quad \text{where } c = \frac{c_2}{c_1}$$

Or
$$\frac{1}{c_1} [\ln(T(t) - c)]_{T_o}^{T(t)} = -\frac{\tau}{\rho L c_P}$$

Or
$$\frac{1}{c_1} [\ln(T(t) - c)]_{T(t)}^{T_o} = \frac{\tau}{\rho L c_P}$$

Or
$$\tau = \frac{\rho L c_P}{c_1} \ln \left[\frac{T_o - c}{T(t) - c} \right] \quad (1)$$

$$c_1 = h_w + h_a = 100 + 20 = 120$$

$$c_2 = h_w T_w + h_a T_a = 100 \times 30 + 20 \times 30 = 3600$$

$$c = \frac{c_2}{c_1} = \frac{3600}{120} = 30$$

Substituting the proper values in equation (1), we get:

$$\tau = \frac{8800 \times 0.02 \times 400}{120} \ln\left[\frac{150-30}{90-30}\right] = 406.6 \text{ s or } 6.776 \text{ minutes}$$

7.3 Example (3): Determining the Conditions under which the Contact Surface Remains at Constant Temperature

Two infinite bodies of thermal conductivities k_1 and k_2, thermal diffusivities α_1 and α_2 are initially at temperatures t_1 and t_2 respectively. Each body has single plane surface and these surfaces are placed in contact with each other. Determine the conditions under which the contact surface remains at constant temperature t_s where $t_1 > t_s > t_2$.

Solution:

The rate of heat flow at a surface ($x = 0$) is given by,

$$Q = \frac{-kA\Delta t}{\sqrt{\pi \alpha \tau}}$$

Heat received by each unit area of contact surface from the body at temperature t_1 is,

$$Q_1 = \frac{-k_1(t_1 - t_s)}{\sqrt{\pi \alpha_1 \tau}}$$

Heat lost by each unit area of contact surface from the body at temperature t_2 is,

$$Q_2 = \frac{-k_2(t_s - t_2)}{\sqrt{\pi \alpha_2 \tau}}$$

The contact surface will remain at a constant temperature if:

$$\frac{-k_1(t_1 - t_s)}{\sqrt{\pi \alpha_1 \tau}} = \frac{-k_2(t_s - t_2)}{\sqrt{\pi \alpha_2 \tau}}$$

Or

$$\frac{k_1(t_1 - t_s)}{\sqrt{\alpha_1}} = \frac{k_2(t_s - t_2)}{\sqrt{\alpha_2}}$$

Or

$$k_1(t_1 - t_s)\sqrt{\alpha_2} = k_2(t_s - t_2)\sqrt{\alpha_1}$$

Or
$$k_1 t_1 \sqrt{\alpha_2} - k_1 t_s \sqrt{\alpha_2} = k_2 t_s \sqrt{\alpha_1} - k_2 t_2 \sqrt{\alpha_1}$$

Or
$$t_s(k_1 \sqrt{\alpha_2} + k_2 \sqrt{\alpha_1}) = k_1 t_1 \sqrt{\alpha_2} + k_2 t_2 \sqrt{\alpha_1}$$

Or
$$t_s = \frac{k_1 t_1 \sqrt{\alpha_2} + k_2 t_2 \sqrt{\alpha_1}}{k_1 \sqrt{\alpha_2} + k_2 \sqrt{\alpha_1}}$$

By dividing the numerator and the denominator by $\sqrt{\alpha_1 \alpha_2}$, the following formula is obtained:

$$t_s = \frac{(k_1 t_1/\sqrt{\alpha_1}) + (k_2 t_2/\sqrt{\alpha_2})}{(k_1/\sqrt{\alpha_1}) + (k_2/\sqrt{\alpha_2})}$$

7.4 Example (4): Calculation of the Time Required for the Plate to Reach a Given Temperature

A $50cm \times 50cm$ copper slab 6.25mm thick has a uniform temperature of $300°C$. Its temperature is suddenly lowered to $36°C$. Calculate the time required for the plate to reach the temperature of $108°C$.

Take: $\rho = 9000 kg/m^3$; $c_P = 0.38 kJ/kg°C$; $k = 370 w/m°C$ and $h = 90 w/m^{2°}C$

Solution:

Surface area of plate (two sides),
$$A_s = 2 \times 0.5 \times 0.5 = 0.5 m^2$$

Volume of plate,
$$V = 0.5 \times 0.5 \times 0.00625 = 0.0015625 m^3$$

Characteristic length,
$$L = \frac{V}{A_s} = \frac{0.0015625}{0.5} = 0.003125 \, m$$

$$Bi = \frac{hL}{k} = \frac{90 \times 0.003125}{370} = 7.6 \times 10^{-4}$$

Since, $Bi \ll 0.1$, hence lumped capacitance method (Newtonian heating or cooling) may be applied for the solution of the problem.

The temperature distribution is given by:

$$\frac{\theta}{\theta_o} = \frac{T(t) - T_\infty}{T_o - T_\infty} = e^{-Bi \times Fo}$$

$$Fo = \frac{k}{\rho c_p L^2} \cdot \tau = \frac{370}{9000 \times 0.38 \times 10^3 \times 0.003125^2} \cdot \tau = 11.0784\, \tau$$

$$\frac{\theta}{\theta_o} = \frac{108 - 36}{300 - 36} = e^{-7.6 \times 10^{-4} \times 11.0784\, \tau} = e^{-8.42 \times 10^{-3}\, \tau}$$

$$0.27273 = e^{-8.42 \times 10^{-3}\, \tau}$$

$$\ln 0.27273 = -8.42 \times 10^{-3}\, \tau \, \ln e$$

$$\tau = \frac{\ln 0.27273}{-8.42 \times 10^{-3} \, \ln e} = 154.31\, s$$

7.5 Example (5): Determination of the Time Required for the Plate to Reach a Given Temperature

An aluminum alloy plate of $400mm \times 400mm \times 4mm$ size at $200°C$ is suddenly quenched into liquid oxygen at $-183°C$. Starting from fundamentals or deriving the necessary expression, determine the time required for the plate to reach a temperature of $-70°C$. Assume $h = 20000 kJ/m^2 \cdot hr \cdot °C$

$c_p = 0.8 kJ/kg°C$, and $\rho = 3000 kg/m^3$, k for aluminum at low temperature may be taken as $214 w/m°C$ or $770.4\, kJ/mh°C$

Solution:

Surface area of the plate,

$$A_s = 2 \times 0.4 \times 0.4 = 0.32 m^2$$

Volume of the plate,

$$V = 0.4 \times 0.4 \times 0.004 = 0.00064 m^3$$

Characteristic length,

$$L = \frac{t}{2} = \frac{0.004}{2} = 0.002 \, m$$

Or

$$L = \frac{V}{A_s} = \frac{0.00064}{0.32} = 0.002 \, m$$

k for aluminum, at low temperature may be taken as $214 w/m°C$ or $770.4 kJ/mh°c$

$$\therefore Bi = \frac{hL}{k} = \frac{2000 \times 0.002}{770.4} = 0.0519$$

Since, $Bi \ll 0.1$, hence lumped capacitance method may be applied for the solution of the problem.

The temperature distribution is given by:

$$\frac{\theta}{\theta_o} = \frac{T(t) - T_\infty}{T_o - T_\infty} = e^{-Bi \times Fo}$$

$$Fo = \frac{k}{\rho c_p L^2} \cdot \tau = \frac{214}{3000 \times 0.8 \times 10^3 \times 0.002^2} \cdot \tau = 22.3 \, \tau$$

$$Bi \times Fo = 0.0519 \times 22.3 = 1.15737 \, \tau$$

$$\therefore \frac{\theta}{\theta_o} = \frac{-70 - (-183)}{200 - (-183)} = = e^{-1.15737 \, \tau}$$

$$\frac{113}{383} = e^{-1.15737 \, \tau}$$

$$0.295 = e^{-1.15737 \, \tau}$$

$$\ln 0.295 = -1.15737 \, \tau \ln e$$

$$\tau = \frac{\ln 0.295}{-1.15737 \times 1} = 1.055 \, s$$

7.6 Example (6): Determining the Temperature of a Solid Copper Sphere at a Given Time after the Immersion in a Well – Stirred Fluid

A solid copper sphere of $10 cm$ diameter ($\rho = 8954 kg/m^3$, $c_p = 383 J/kg$, $K = 386 w/mK$) initially at a uniform temperature of $T_o = 250°C$, is suddenly immersed in a well – stirred fluid which is maintained at a uniform temperature $T_\infty = 50°C$. The heat transfer coefficient between the sphere and the fluid is $h = 200 w/m^2 K$. Determine the temperature of the copper block at $\tau = 5\ min.$ after the immersion.

Solution:

Given: $d = 10cm = 0.1m$; $\rho = 8954 kg/m^3$; $c_p = 383 J/kg\ k$; $k = 386 w/m\ K$, $T_o = 250°C$; $T_\infty = 50°C$; $h = 200 w/m^2 K$; $\tau = 5 min = 5 \times 60 = 300\ s$

Temperature of the copper block, $T(t) = ?$

The characteristic length of the sphere is,

$$L = \frac{V}{A_s} = \frac{\frac{4}{3}\pi r^3}{4\pi r^2} = \frac{r}{3} = \frac{d}{6} = \frac{0.1}{6} = 0.01667 m$$

$$Bi = \frac{hL}{k} = \frac{200 \times 0.01667}{386} = 8.64 \times 10^{-3}$$

Since, $Bi \ll 0.1$, hence lumped capacitance method (Newtonian heating or cooling) may be applied for the solution of the problem.

The temperature distribution is given by:

$$\frac{\theta}{\theta_o} = \frac{T(t) - T_\infty}{T_o - T_\infty} = e^{-Bi \times Fo}$$

$$Fo = \frac{k}{\rho c_p L^2} \cdot \tau = \frac{386}{8954 \times 383 \times 0.01667^2 \times 300} = 121.513$$

$$Bi \times Fo = 8.54 \times 10^{-3} \times 121.513 = 1.05$$

$$\therefore \frac{\theta}{\theta_o} = \frac{-T(t) - 50}{250 - 50} = e^{-1.05}$$

$$T(t) - 50 = 200\ e^{-1.05}$$

$$\therefore T(t) = 50 + 200\ e^{-1.05} = 50 + 70 = 120°c$$

7.7 Example (7): Determination of the Heat Transfer Coefficient

An average convective heat transfer coefficient for flow of $90°C$ air over a plate is measured by observing the temperature – time history of a $40mm$ thick copper slab ($\rho = 9000 kg/m^3, c_p = 0.38 kj/kg°C, k = 370 w/m°C$) exposed to $90°C$ air. In one test run, the initial temperature of the plate was $200°C$, and in 4.5 minutes the temperature decreased by $35°C$. Find the heat transfer coefficient for this case. Neglect internal thermal resistance.

Solution:

Given: $T_\infty = 90°C; t = 40mm$ or $0.04m; \rho = 9000 kg/m^3; c_p = 0.38 kj/kg°C;$
$T_o = 200°C; T(t) = 200 - 35 = 165°C; \tau = 4.5 min = 270s$

Characteristic length of the sphere is,

$$L = \frac{t}{2} = \frac{0.04}{2} = 0.02 m$$

$$Bi = \frac{hL}{k} = \frac{0.02h}{370} = 5.405 \times 10^{-5} h$$

$$Fo = \frac{k}{\rho c_p L^2} \cdot \tau = \frac{370}{9000 \times 0.38 \times 10^3 \times 0.02^2} \times 270 = 73.03$$

$$Bi \times Fo = 5.405 \times 10^{-5} \times 73.03 h = 394.73 \times 10^{-5} h = 0.003947 h$$

$$\frac{\theta}{\theta_o} = \frac{T(t) - T_\infty}{T_o - T_\infty} = e^{-Bi \times Fo}$$

$$\frac{\theta}{\theta_o} = \frac{165 - 90}{200 - 90} = e^{-0.003947 h}$$

$$0.682 = e^{-0.003947 h}$$

$$\ln 0.682 = -0.003947 h \ln e$$

$$h = \frac{\ln 0.682}{-0.003947 \times 1} = 96.97 w/m^{2°}C$$

7.8 Example (8): Determination of the Heat Transfer Coefficient

The heat transfer coefficients for flow of air at $28°C$ over a $12.5mm$ diameter sphere are measured by observing the temperature – time history of a copper ball of the same dimension. The temperature of the copper ball ($c_p = 0.4 kj/kg\ K$ and $\rho = 8850 kg/m^3$) was measured by two thermo – couples, one located in the center and the other near the surface. Both the thermocouples registered the same temperature at a given instant. In one test the initial temperature of the ball was $65°C$, and in 1.15 minutes the temperature decreased by $11°C$. Calculate the heat transfer coefficient for this case.

Solution:

Given: $T_\infty = 28°C; r(\text{sphere}) = \frac{12.5}{2} = 6.25mm = 0.00625m;\ c_p = 0.4 kj/kg\ K;$

$\rho = 8850 kg/m^3;\ T_o = 65°C;\ T(t) = 65 - 11 = 54°C;\ \tau = 1.15 min = 69\ s;$

$Bi = \frac{hL}{k}$; characteristic length $L = \frac{V}{A_s} = \frac{r}{3} = \frac{0.00625}{3} = 2.083 \times 10^{-3}$

Since, heat transfer coefficient has to be calculated, so assume that the internal resistance is negligible and Bi is much less than 0.1.

$$Bi = \frac{hr}{3k} = \frac{0.00625h}{3k} = \frac{2.083 \times 10^{-3} h}{k}$$

$$Fo = \frac{k}{\rho c_p L^2} \cdot \tau = \frac{k}{8850 \times 0.4 \times 10^3 \times (2.083 \times 10^{-3})^2} \times 270$$

$$= 0.0651\ k \times 69 = 4.5\ k$$

$$Bi \times Fo = \frac{0.00625h}{3k} \times 4.5k = 9.375 \times 10^{-3} h$$

$$\frac{\theta}{\theta_o} = \frac{T(t) - T_\infty}{T_o - T_\infty} = e^{-Bi \times Fo}$$

$$\frac{\theta}{\theta_o} = \frac{54 - 28}{65 - 28} = e^{-9.375 \times 10^{-3} h}$$

$$0.7027 = e^{-0.009375h}$$

$$\ln 0.7027 = -0.009375 h \ln e$$

$$\therefore h = \frac{\ln 0.7027}{-0.009375 \times 1} = 37.63 w/m^2 K$$

7.9 Example (9): Calculation of the Initial Rate of Cooling of a Steel Ball

A steel ball 50mm in diameter and at $900°C$ is placed in still atmosphere of $30°C$. Calculate the initial rate of cooling of the ball in $°C/min$, if the duration of cooling is 1 minute.

Take: $\rho = 7800 kg/m^3$; $c_p = 2kj/kg°C$ (for steel); $h = 30w/m^{2°}C$.
Neglect internal thermal resistance.

Solution:

Given: $r = \frac{50}{2} = 25mm = 0.025m$; $T_o = 900°C$; $T_\infty = 30°C$; $\rho = 7800 kg/m^3$
$c_p = 2kj/kg°C$; $h = 30w/m^{2°}C$; $\tau = 1 min = 60 s$;

The temperature variation in the ball (with respect to time), neglecting internal thermal resistance, is given by:

$$\frac{\theta}{\theta_o} = \frac{T(t) - T_\infty}{T_o - T_\infty} = e^{-Bi \times Fo}$$

$$Bi = \frac{hL}{k}, \quad L \text{ of a ball} = \frac{r}{3} = \frac{0.025}{3}$$

$$\therefore Bi = \frac{hr}{3k} = \frac{30 \times 0.025}{3k} = \frac{0.25}{k}$$

$$Fo = \frac{k}{\rho c_p L^2} \cdot \tau = \frac{k}{7800 \times 2 \times 10^3 \times (0.025/3)^2} \cdot \tau$$

$$Fo = 9.23 \times 10^{-4} \times 60 = 0.0554 k$$

$$Bi \times Fo = \frac{0.25}{k} \times 0.0554 k = 0.01385$$

$$\frac{\theta}{\theta_o} = \frac{T(t) - 30}{900 - 30} = e^{-0.01385}$$

$$T(t) = 870\, e^{-0.01385} + 30 = 888°C$$

$$\therefore \text{rate of cooling} = \frac{900 - 888}{1\, min} = 12°C/min$$

7.10 Example (10): Determination of the Maximum Speed of a Cylindrical Ingot inside a Furnace

A cylinder ingot 10cm diameter and 30cm long passes through a heat treatment furnace which is 6m in length. The ingot must reach a temperature of $800°C$ before it comes out of the furnace. The furnace gas is at $1250°C$ and the ingot initial temperature is $90°C$. What is the maximum speed with which the ingot should move in the furnace to attain the required temperature? The combined radiative and convective surface heat transfer coefficient is $100w/m^{2°}C$.

Take: k (steel) = $40w/m^{2°}C$ and α (thermal diffusivity of steel) = $1.16 \times 10^{-5} m^2/s$.

Solution:

$d = 10cm = 0.1m$; $L = 30cm = 0.3m$; Length of the furnace = $6m$;

$T_o = 800°C$; $T(t) = 800°C$; $T_\infty = 90°C$;

v_{max} of ingot passing through the furnace =?

$h = 100w/m^{2°}C$; $k(steel) = 40w/m°C$; $\alpha(steel) = 1.16 \times 10^{-5} m^2/s$

Characteristic length,

$$L_c = \frac{V}{A_s} = \frac{\frac{4}{3}d^2 L}{\pi dL + \frac{\pi}{4}d^2 \times 2} = \frac{dL}{4L + 2d}$$

$$= \frac{0.1 \times 0.3}{4 \times 0.3 + 2 \times 0.1} = 0.02143m$$

$$Bi = \frac{hL_c}{k} = \frac{100 \times 0.02143}{40} = 0.0536$$

As $Bi \ll 0.1$, Then internal thermal resistance of the ingot for conduction heat flow can be neglected.

∴ The time versus temperature relation is given as:

$$\frac{T(t) - T_\infty}{T_o - T_\infty} = e^{-Bi \times Fo}$$

$$Fo = \frac{k}{\rho c_p L^2} \cdot \tau = \frac{\alpha}{L^2} \cdot \tau = \frac{1.16 \times 10^{-5}}{0.02143^2} \cdot \tau = 0.02526\,\tau$$

$$Bi \times Fo = 0.0536 \times 0.02526\,\tau = 1.35410^{-3}\tau = 0.001354\,\tau$$

$$\frac{\theta}{\theta_o} = \frac{T(t) - T_\infty}{T_o - T_\infty} = e^{-Bi \times Fo}$$

$$\frac{\theta}{\theta_o} = \frac{800 - 90}{1250 - 90} = e^{-0.001354\,\tau}$$

$$0.6121 = e^{-0.001354\,\tau}$$

$$-0.001354\,\tau \ln e = \ln 0.6121$$

$$\tau = \frac{\ln 0.6121}{-0.001354} = 362.5\,s$$

v_{max} of ingot passing through the furnace,

$$v_{max} = \frac{\text{furnace length}}{\text{time}} = \frac{6}{362.5} = 0.01655\,m/s$$

7.11 Example (11): Determining the Time Required to Cool a Mild Steel Sphere, the Instantaneous Heat Transfer Rate, and the Total Energy Transfer

A $15mm$ diameter mild steel sphere ($k = 4.2w/m°C$ is exposed to cooling air flow at $20°c$ resulting in the convective, coefficient $h = 120w/m^{2°}C$.

Determine the following:

(i) Time required to cool the sphere from $550°C$ to $90°C$.

(ii) Instantaneous heat transfer rate 2 minutes after the start of cooling.

(iii) Total energy transferred from the sphere during the first 2 minutes.

For mild steel take: $\rho = 7850 kg/m^3$; $c_p = 475 J/kg°C$; and $\alpha = 0.045 \, m^2/h$

Solution:

Given: $r = \frac{15}{2} = 7.5mm = 0.0075m$; $k = 42 w/m^{2°}C$; $T_\infty = 20°C$; $T_o = 550°C$;

$T(t) = 90°C$; $h = 120 w/m^{2°}C$;

(i) Time required to cool the sphere from $550°C$ to $90°C, \tau = ?$

The characteristic length, L_c is given by,

$$L_c = \frac{r}{3} = \frac{0.0075}{3} = 0.0025m$$

Biot number,

$$Bi = \frac{hL_c}{k} = \frac{120 \times 0.0025}{42} = 0.007143$$

Since, $Bi \ll 0.1$, so we can use the lumped capacitance theory to solve this problem.

Fourier Number,

$$Fo = \frac{k}{\rho c_p L^2} \cdot \tau = \frac{\alpha}{L_c^2} \cdot \tau$$

$$\alpha = 0.045 m^2/h = \frac{0.045}{3600} = 1.25 \times 10^{-5} m^2/s$$

$$Fo = \frac{1.25 \times 10^{-5}}{(0.0025)^2} \cdot \tau = 2\tau$$

$$Bi \times Fo = 0.007143 \times 2\tau$$

The temperature variation with time is given by:

$$\frac{\theta}{\theta_o} = \frac{T(t) - T_\infty}{T_o - T_\infty} = e^{-Bi \times Fo}$$

$$= \frac{90 - 20}{550 - 20} = e^{-0.014286 \, \tau}$$

$$0.132 = e^{-0.014286\,\tau}$$

$$-0.014286\,\tau \ln e = \ln 0.132$$

$$\tau = \frac{\ln 0.132}{-0.014286} = 141.7\ s$$

(ii) Instantaneous heat transfer rate 2 minutes (0.0333h) after the start of cooling, $q'(\tau) = ?$

$$q'(\tau) = hA_s\theta_o e^{-Bi \times Fo}$$

$$Bi \times Fo = 0.014286 \times 2 \times 60 = 1.7143$$

$$q'(\tau) = 120 \times 4\pi \times (0.0075)^2 (550 - 20) e^{-1.7143} = 8.1 w$$

(iii) Total energy transferred from the sphere during the first 2 minutes, (0.0333h) $Q(t) = ?$

$$Q(t) = hA_s\theta_o \left[1 - e^{-Bi \times Fo}\right] \frac{\tau}{Bi \times Fo}$$

$$= 120 \times 4\pi \times (0.0075)^2 (550 - 20)[1 - e^{-1.7143}] \times \frac{120}{1.7143} = 2580.2\ J$$

Or

$$Fo = \frac{k}{\rho c_p L_c^2} \cdot \tau = \frac{k}{7850 \times 475 \times 0.0025^2} \times 120 = 206$$

$$Bi \times Fo = 0.007143 \times 206 = 1.471$$

$$Q(t) = 120 \times 4\pi \times (0.0075)^2 (550 - 20)[1 - e^{-1.471}] \times \frac{120}{1.471} = 2825\ J$$

7.12 Example (12): Estimation of the Time Required to Cool a Decorative Plastic Film on Copper Sphere to a Given Temperature using Lumped Capacitance Theory

The decorative plastic film on copper sphere 10mm in diameter is cured in an oven at 75°C. After removal from oven, the sphere is exposed to an air stream at 10m/s and 23°C. Estimate the time taken to cool the sphere to 35°C using lumped capacitance theory.

Use the following correlation:

$$Nu = 2 + \left[0.4(Re)^{0.5} + 0.06(Re)^{2/3}\right](pr)^{0.4} \left[\frac{\mu_a}{\mu_s}\right]^{0.25}$$

For determination of correlation coefficient h, use the following properties of air and copper:

For copper: $\rho = 8933 kg/m^3$; $k = 400 w/m\ K$; $c_p = 380 J/kg°C$

For air at $23°C$: $\mu_a = 18.6 \times 10^{-6} Ns/m^2$, $v = 15.36 \times 10^{-6} m^2/s$

$$k = 0.0258 w/m\ K, \qquad pr = 0.709, \text{ and}$$
$$\mu_s = 19.7 \times 10^{-6} Ns/m^2, \qquad \text{at } 35°C$$

Solution:

$d = 10mm = 0.01m$; $T_o = 75°C$; $v = 10m/s$; $T_\infty = 23°C$; $T(t) = 35°C$;

Time taken to cool the sphere, $\tau = ?$

$$Re = \frac{v\ d}{v} = \frac{10 \times 0.01}{15.36 \times 10^{-6}} = 6510$$

$$Nu = 2 + \left[0.4(6510)^{0.5} + 0.06(6510)^{2/3}\right] \times (0.709)^{0.4} \times \left[\frac{18.16 \times 10^{-6}}{19.78 \times 10^{-6}}\right]^{0.25}$$

$$= 2 + [32.27 + 20.92] \times 0.87 \times 0.979 = 47.3$$

$$\text{or} \quad Nu = \frac{h\ d}{k} = 47.3$$

$$\therefore h = \frac{k}{d} Nu = \frac{0.0258}{0.01} \times 47.3 = 122 w/m^{2°}C$$

The time taken to cool from $75°C$ to $35°C$ may be found from the following relation:

$$\frac{\theta}{\theta_o} = \frac{T(t) - T_\infty}{T_o - T_\infty} = e^{-Bi \times Fo}$$

$$Bi = \frac{hL}{k}$$

The characteristic length of a sphere, $L = \frac{r}{3} = \frac{0.005}{3} m$

$$Bi = \frac{hL}{k} = \frac{122 \times 0.005}{3 \times 400} = 5.083 \times 10^{-4}$$

Since, $Bi \ll 0.1$, so we can use the lumped capacitance theory to solve this problem.

$$Fo = \frac{k}{\rho c_p L^2} \cdot \tau = \frac{400}{8933 \times 380 \times \left(\frac{0.005}{3}\right)^2} \cdot \tau = 42.421\, \tau$$

$$Bi \times Fo = 5.083 \times 10^{-4} \times 42.421\, \tau = 0.0216\, \tau$$

$$\frac{\theta}{\theta_o} = \frac{35 - 23}{75 - 23} = e^{-0.0216\, \tau}$$

$$0.2308 = e^{-0.0216\, \tau}$$

$$\ln 0.2308 = -0.0216\, \tau \ln e$$

$$\tau = \frac{\ln 0.2308}{-0.0216} = 67.9 \simeq 68\, s$$

7.13 Example (13): Calculation of the Time Taken to Boil an Egg

An egg with mean diameter of 40mm and initially at $20°C$ is placed in a boiling water pan for 4 minutes and found to be boiled to the consumer's taste. For how long should a similar egg for the same consumer be boiled when taken from a refrigerator at $5°C$. Take the following properties for eggs:

$k = 10 w/m°C;\ \rho = 1200 kg/m^3;\ c_p = 2 kj/kg°C;$ and

h (heat transfer coefficient) $= 100 w/m^{2°}C$.

Use lumped capacitance theory.

Solution:

Given: $r = \frac{40}{2} = 20mm = 0.02m$; $T_o = 20°c$; $T_\infty = 100°C$; $\tau = 4 \times 60 = 240s$;

$k = 10w/m°C$; $\rho = 1200 kg/m^3$; $c_p = 2kj/kg°C$; $h = 100w/m^{2°}C$;

For using the lumped capacitance theory, the required condition $Bi \ll 0.1$ must be valid.

$Bi = \frac{hL}{k}$, where L is the characteristic length which is given by,

$$L = \frac{V}{A_s} = \frac{r}{3} = \frac{0.02}{3} m$$

$$\therefore Bi = \frac{hL}{k} = \frac{h \times 0.02}{k \times 3} = \frac{100 \times 0.02}{10 \times 3} = 0.067$$

As $Bi \ll 0.1$, we can use the lumped capacitance system.

The temperature variation with time is given by:

$$\frac{T(t) - T_\infty}{T_o - T_\infty} = e^{-Bi \times Fo}$$

$$Fo = \frac{k}{\rho c_p L^2} \cdot \tau = \frac{10}{1200 \times 2 \times 10^3 \times \left(\frac{0.02}{3}\right)^2} \times 240 = 22.5$$

$$Bi \times Fo = 0.067 \times 22.5 = 1.5075$$

$$\frac{T(t) - 100}{20 - 100} = e^{-1.5075}$$

$$T(t) = 100 - 80\, e^{-1.5075} = 82.3°C \simeq 82°C$$

Now, let us find τ when the given data is: $T_o = 5°C$; $T_\infty = 100°C$ and $T(t) = 82°C$.

$$\frac{82 - 100}{5 - 100} = e^{-Bi \times Fo}$$

$$Fo = \frac{k}{\rho c_p L^2} \cdot \tau = \frac{10}{1200 \times 2 \times 10^3 \times \left(\frac{0.02}{3}\right)^2} \cdot \tau = 0.09375\, \tau$$

$$Bi \times Fo = 0.067 \times 0.09375\,\tau = 6.281 \times 10^{-3}\tau = 0.00628\,\tau$$

$$0.1895 = e^{-0.00628\,\tau}$$

$$-0.00628\,\tau \ln e = \ln 0.1895$$

$$\tau = \frac{\ln 0.1895}{-0.00628} = 264.9\,s = 4.414\,minutes$$

7.14 Example (14): Determining the Total Time Required for a Cylindrical Ingot to be heated to a Given Temperature

A hot cylinder ingot of 50mm diameter and 200mm long is taken out from the furnace at $800°C$ and dipped in water till its temperature fall to $500°C$. Then, it is directly exposed to air till its temperature falls to $100°C$. Find the total time required for the ingot to reach the temperature from $800°C$ to $100°C$. Take the following:

k(thermal conductivity of ingot) $= 60w/m°C$;
c(specific heat of ingot) $= 200J/m°C$;
ρ(density of ingot material) $= 800kg/m^3$;
h_w(heat transfer coefficient in water) $= 200w/m^{2°}C$;
h_a(heat transfer coefficient in air) $= 20w/m^{2°}C$;
Temperature of air or water $= 30°C$

Solution:

Given: $r = \frac{50}{2} = 25mm = 0.025m; L = 200mm$ or $0.2m$

The characteristic length of a cylinder,

$$L_c = \frac{r}{2} = \frac{0.025}{2} = 0.0125m$$

$$Bi = \frac{hL}{k} = \frac{200 \times 0.0125}{60} = 0.04167$$

As Bi is less than 0.1, the internal thermal resistance can be neglected, and lumped capacitance theory can be used. The total time (τ) can be calculated by calculating τ_1 (time required in water) and τ_2 (time required in air) and adding them such that $\tau = \tau_1 + \tau_2$

(a) The temperature variation with respect to time when cooled in water is given by: (see Fig. (7.2) below)

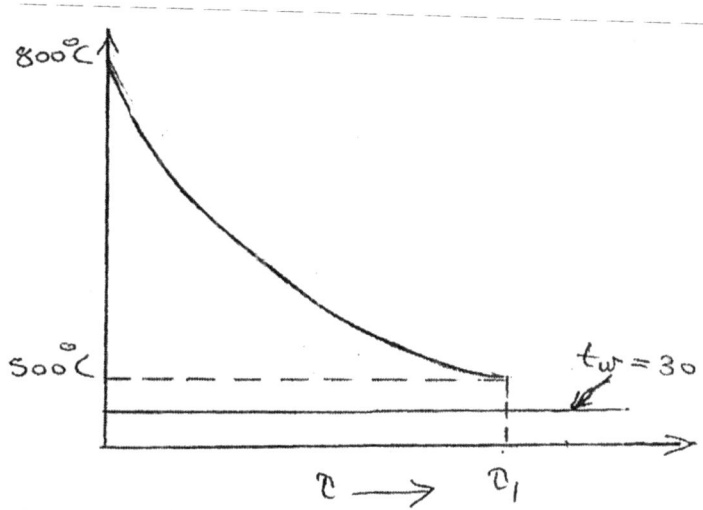

Fig. (7.2) Temperature variation with time when the ingot is cooled in water

$$\frac{T(t) - T_w}{T_o - T_w} = e^{-Bi \times Fo}$$

$$Fo = \frac{k}{\rho c_p L^2} \cdot \tau_1 = \frac{60}{800 \times 200 \times (0.0125)^2} \cdot \tau_1 = 2.4\,\tau_1$$

$$Bi \times Fo = 0.04167 \times 2.4\,\tau_1 = 0.1\,\tau_1$$

$$\therefore \frac{\theta}{\theta_o} = \frac{T(t) - T_w}{T_o - T_w} = e^{-Bi \times Fo}$$

$$= \frac{500 - 30}{800 - 30} = e^{-0.1\,\tau_1}$$

$$0.61 = e^{-0.1\,\tau_1}$$

$$-0.1\,\tau_1\,\ln e = \ln 0.61$$

$$\tau_1 = \frac{\ln 0.61}{-0.1} = 4.94\ s$$

(b) The temperature variation with respect to time when cooled in air is given by: (see Fig. (7.3) below)

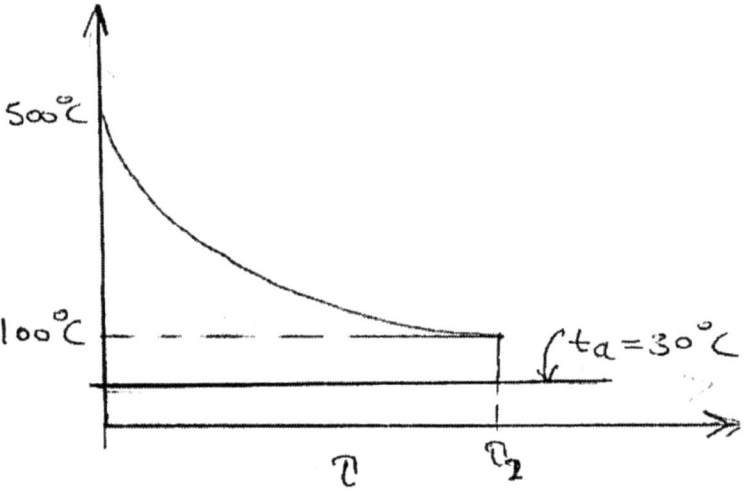

Fig. (7.3) Temperature variation with time when the ingot is cooled in air

$$\therefore \frac{\theta}{\theta_o} = \frac{T(t) - T_a}{T_o - T_a} = e^{-Bi \times Fo}$$

$$Bi = \frac{hL}{k} = \frac{20 \times 0.0125}{60} = 0.004167$$

$$Fo = \frac{k}{\rho c_p L^2} \cdot \tau_2 = 2.4\,\tau_2$$

$$Bi \times Fo = 0.004167 \times 2.4\,\tau_2 = 0.01\,\tau_2$$

$$\frac{\theta}{\theta_o} = \frac{100 - 30}{500 - 30} = \frac{70}{470} = e^{-0.01\,\tau_2}$$

$$0.149 = e^{-0.01\,\tau_2}$$

$$-0.01\, \tau_2\, \ln e = \ln 0.149$$

$$\therefore\ \tau_2 = \frac{\ln 0.149}{-0.01 \times 1} = 190.42\ s$$

∴ Total time (τ) is given by:

$$\tau = \tau_1 + \tau_2 = 4.94 + 190.42 = 195.36\ s = 3.256\ min.$$

Chapter Eight

Unsolved Theoretical Questions and Further Problems in Lumped Capacitance System

8.1 Theoretical Questions

1. What is meant by transient heat conduction?
2. What is lumped capacity?
3. What are the assumptions for lumped capacity analysis?
4. What are Fourier and Biot numbers? What is the physical significance of these numbers?
5. Define a semi – infinite body.
6. What is an error function? Explain its significance in a semi – infinite body in transient state.
7. What are Heisler charts?
8. Explain the significance of Heisler charts in solving transient conduction problems.

8.2 Further Problems

1. A copper slab ($\rho = 9000 kg/m^3, c = 380 J/kg°C, k = 370 w/m°C$) measuring $400mm \times 400mm \times 5mm$ has a uniform temperature of $250°c$. Its temperature is suddenly lowered to $30°C$. Calculate the time required for the plate to reach the temperature of $90°C$. Assume convective heat transfer coefficient as $90 w/m^{2°}C$.

Ans. $\{123.75\ s\}$

2. An aluminum alloy plate $0.2 m^2$ surface area (both sides), 4mm thick and at $200°C$ is suddenly quenched into liquid oxygen which is at $-183°C$. Find the time required for the plate to reach the temperature of $-70°C$.

Take: $\rho = 2700 kg/m^3, c = 890 J/kg°C, h = 500 w/m^{2°}C$

Ans. $\{23.45\ s\ \}$

3. A sphere of 200mm diameter made of cast iron initially at a uniform temperature of $400°C$ is quenched into oil. The oil bath temperature is $40°C$. If the temperature of sphere is $100°C$ after 5 minutes, find heat transfer coefficient on the surface of the sphere.

Take: c_P (cast iron) $=0.32 KJ/kg°C$; ρ(cast iron) $=7000 kg/m^3$

Neglect internal thermal resistance.

$$Ans. \{134 kw/m^{2°}C\}$$

4. An average convective heat transfer coefficient for flow of $100°C$ air over a flat plate is measured by observing the temperature – time history of a 30 mm thick copper slab ($\rho = 9000 kg/m^3, c = 0.38 kJ/kg°C, k = 370 w/m°C$) exposed to $100°C$ air. In one test run, the initial temperature of the plate was $210°C$ and in 5 minutes the temperature decreased by $40°C$. Find the heat transfer coefficient for this case. Neglect internal thermal resistance.

$$Ans. \{77.24 w/m^{2°}C\}$$

5. A cylinder steel ingot 150mm in diameter and 400mm long passes through a heat treatment furnace which is 6m in length. The ingot must reach a temperature of $850°C$ before it comes out of the furnace. The furnace gas is at $1280°C$ and ingot initial temperature is $100°C$. What is the maximum speed with which the ingot should move in the furnace to attain the required temperature? The combined radiative and convective surface heat transfer coefficient is $100 w/m^{2°}C$. Take k (steel)= $45 w/m°C$ and α (thermal diffusivity) $= 0.46 \times 10^{-5} m^2/s$.

$$Ans. \{ 1.619 \times 10^{-3} m/s \}$$

6. A hot mild steel sphere ($k = 42.5 w/m°C$) having 12mm diameter is planned to be cooled by an air flow at $27°C$. The convective heat transfer coefficient is $114 w/m^{2°}C$. Determine the following:

(i) Time required to cool the sphere from $540°C$ to $95°C$;
(ii) Instantaneous heat transfer rate 2 minutes after the start of cooling;
(iii) Total energy transferred from the sphere during the first 2 minutes. Take mild steel properties as ($\rho = 7850 kg/m^3, c = 475 kJ/kg°C, \alpha = 0.043 m^2/h$).

$$Ans. \{ (i)\ 2.104\ min;\ (ii)\ 3.884 w\ ;\ (iii)\ 1475.7\ J \}$$

7. The heat transfer coefficients for the flow of $30°C$ air over a 12.5mm diameter sphere are measured from observing the temperature – time history of a copper ball of the same dimensions. The temperature of the copper ball ($\rho = 8930 kg/m^3$; $c = 0.375 kJ/kg°C$) was measured by two thermocouples, one located at the center and the other near the surface. Both thermocouples registered within the accuracy of the recording instruments the same temperature at the given instant, on one test run, the initial temperature of the ball was $70°C$ and in 1.15 minutes the temperature decreased by $7°C$. Calculate the convective heat transfer coefficient for this case.

$$\text{Ans.} \{194.5 \ w/m^{2°}C\}$$

8. The temperature of an air stream flowing with a velocity of 3 m/s is measured by a copper – constantan thermocouple which may be approximated as a sphere 3mm in diameter. Initially the junction and air are at a temperature of $25°C$. The air temperature suddenly changes to and is maintained at $200°C$.

i) Determine the time required for the thermocouple to indicate a temperature of $150°C$. Also determine the thermal time constant and temperature indicated by the thermocouple at that instant;

ii) Discuss the suitability of this thermocouple to measure unsteady state temperature of a fluid when the temperature variation in the fluid has a time period of 3 seconds.

The thermocouple junction properties are:

Density $= 8685 kg/m^3$; specific heat $c = 383 \ j/kg°C$; thermal conductivity (thermocouple) $k = 29 w/m°C$ and convective coefficient $h = 150 w/m^{2°}C$.

$$\text{Ans.} \{13.89 \ s; \ 11.09 \ s; \ 155.63°C\}$$

9. A 50 mm thick large steel plate ($k = 42.5 w/m°C$, $\alpha = 0.043 \ m^2/h$), initially at $425°C$ is suddenly exposed on both sides to an environment with convective heat transfer coefficient $285 w/m^{2°}C$ and temperature $65°C$. Determine the center line temperature and temperature inside the plate 12.5mm from the mid-plane after 3 minutes.

10. A long cylindrical bar ($k = 17.5 w/m°C$, $\alpha = 0.0185\, m^2/h$) of radius 75mm comes out of oven at $815°C$ throughout and is cooled by quenching it in a large bath of $38°C$ coolant. The surface coefficient of heat transfer between the bar surface and the coolant is $175 w/m^{2°}C$. Determine:

(i) The time taken by the shaft to reach $116°C$;

(ii) The surface temperature of the shaft when its center temperature is $116°C$. Also calculate the temperature gradient at the outside surface at the same instant of time.

$$\text{Ans. } \{(i)\; 2102s;\; (ii)\; 92.6°C;\; 546°C/m\}$$

11. A concrete highway may reach a temperature of $55°C$ on a hot summer's day. Suppose that a stream of water is directed on the highway so that the surface temperature is suddenly lowered to $35°C$. How long will it take to cool the concrete to $45°C$ at a depth of 50mm from the surface?

For concrete take α (thermal diffusivity) $= 1.77 \times 10^{-3} m^2/h$

$$\text{Ans. } \{1.51\, h\}$$

12. It is proposed to bury water pipes underground in wet soil which is initially at $4.5°C$. The temperature of the surface of soil suddenly drops to $-5°C$ and remains at this value for 10 hours. Determine the minimum depth at which the pipes be laid if the surrounding soil temperature is to remain above $0°C$ (without water getting frozen). Assume the soil as semi – infinite soild.

For wet soil take α (thermal diffusivity) $= 2.78 \times 10^{-3} m^2/h$

$$\text{Ans. } \{0.167m\}$$

13. A 50 mm thick mild steel plate ($\alpha = 1.25 \times 10^{-5} m^2/s$) is initially at a temperature of $40°C$. It is suddenly exposed on one side to a fluid which causes the surface temperature to increase to and remain at $90°C$. Determine:

(i) The maximum time that the slab be treated as a semi – infinite body;

(ii) The temperature as the center of the slab one minute after the change in surface temperature.

$$\text{Ans. } \{(i)\; 200s\; ;\; (ii)\; 66°C\}$$

14. The initial uniform temperature of a thick concrete wall ($\alpha = 1.58 \times 10^{-3} m^2/h$; $k = 0.937 w/m°C$) of a jet engine test cell is $21°C$. The surface temperature of the wall suddenly rises to $315°C$ when the combination of exhaust gases from the turbojet and spray of cooling water occurs. Determine:

(i) The temperature at a point 75 mm from the surface after 7.5 hours;

(ii) The instantaneous heat flow rate at the specified plane and at the surface itself at the instant mentioned at (i).

Use the solution for semi – infinite solid.

Ans. {(i) $206°C$; (ii) $-1265.6 w/m^2$; $-1425 w/m^2$}

15. The initial uniform temperature of a large mass of material ($\alpha = 0.41 \, m^2/h$) is $120°C$. The surface of the material is suddenly exposed to and held permanently at $5°C$. Calculate the time required for the temperature gradient at the surface to reach $350°C/m$.

Ans. {206s}

16. A motor car of mass 1500 kg travelling at 80 km/h is brought to rest within a period of 5 seconds when brakes are applied. The braking system consists of 4 brakes with each brake band of $350 \, cm^2$ area; these press against steel drum of equivalent area. The brake lining and the drum surfaces ($k = 55 w/m°C$, $\alpha = 1.24 \times 10^{-5} m^2/s$) are at the same temperature and the heat generated during the stoppage action dissipates by flowing across drums. If the drum surface is treated as semi – infinite plate, calculate the maximum temperature rise.

Ans. {$134.11°C$}

17. During periodic heating and cooling of a thick brick wall, the wall temperature varies sinusoidally. The surface temperature ranges from $25°C$ to $75°C$ during a period of 24 hours. Determine the time lag of the temperature wave corresponding to a point located at 250 mm from the wall surface. The properties of the wall material are: ($\rho = 1620 kg/m^3$, $c = 450 J/kg°C$, $k = 0.62 w/m°C$).

Ans. {$6.24 \, h$}

18. A single cylinder ($\alpha = 0.042 \, m^2/h$ from cylinder material) two – stroke I.C. engine operates at $1500 \, r.p.m.$ Calculate the depth where the temperature wave due to variation of cylinder temperature is damped to 1% of its surface value.

Ans. {**1.775 mm**}

19. The temperature distribution across a large concrete slab ($k = 1.2 w/m°C$, $\alpha = 1.77 \times 10^{-3} m^2/h$) 500 mm thick heated from one side as measured by thermocouples approximates to the relation: $t = 60 - 50x + 12x^2 + 20x^3 - 15x^4$ where t is in $°C$ and x is in meters. Considering an area of $5 m^2$, compute the following:

(i) The heat entering and leaving the slab in unit time;

(ii) The heat energy stored in unit time;

(iii) The rate of temperature change at both sides of the slab; and

(iv) The point where the rate of heating or cooling is maximum.

Ans. {(*i*) **300w, 183 w**; (*ii*) **117w**; (*iii*) **42.48 × 10⁻³°c/h, 69.03 × 10⁻³°c/h**; (*iv*) **0.33m**}

References

1. M. David Burghardt and James A. Harbach, "Engineering Thermodynamics", Fourth Edition, Harper Collins College Publishers, (1993).
2. R. K. Rajput, "Engineering Thermodynamics", Third Edition, SI units version, Laxmi Publications LTD, New Delhi, India (2007).
3. Michael J. Moran and Howard N. Shapiro, "Fundamentals of Engineering Thermodynamics", Fifth Edition, SI units, John Wiley and Sons, Inc., England, (2006).
4. Theodore L. Bergman and et al., "Fundamentals of Heat and Mass Transfer", Seventh Edition, John Wiley and Sons, Inc., England, (2011).
5. Eastop and McConkey, "Applied Thermodynamics for Engineering Technologists", Fifth Edition, Longman Group UK Limited, London, (1993).
6. Osama Mohammed Elmardi, "Solution of Problems in Heat and Mass Transfer", in Arabic language, www.ektab.com, December (2015).
7. Osama Mohammed Elmardi, "Lecture Notes on Thermodynamics I", in Arabic language, Mechanical Engineering Department, Faculty of Engineering and Technology, Nile Valley University, (1998).
8. Osama Mohammed Elmardi, "Lecture Notes on Thermodynamics II", in Arabic language, Mechanical Engineering Department, Faculty of Engineering and Technology, Nile Valley University, (2000).
9. R.K. Rajput, "Heat and Mass Transfer", SI units, S. Chand and Company LTD, New Delhi, (2003).
10. Welty J. R., "Fundamentals of Momentum, Heat and Mass Transfer", 3rd edition, John Wiley, (1984).
11. Croft D.R. and Lilley D.G., "Heat Transfer Calculations using Finite Difference Equations", Pavic Publications, (1986).
12. Incropera F. P. and De Witt D.P., "Fundamentals of Heat and Mass Transfer", 3rd edition, John Wiley, (1990).

13. Rogers G.F.C. and Mayhew Y.R., "Thermodynamics and Transport Properties of Fluids", 4th edition, Basil Blackwell, (1987).
14. Eckert E.R. and Drake R.M., "Analysis of Heat and Mass Transfer", Taylor and Francis, (1971).
15. Kern D.Q., "Process Heat Transfer", McGraw – Hill, (1950).
16. M. J. C. Van Gemert, "Theoretical analysis of the lumped capacitance method in dielectric time domain spectroscopy", Journal of Advances in Molecular Relaxation Processes, Received 12 November (1973), Available online 3 October (2002).
17. Faghri A., Zhang Y., and Howell J. R., "Advanced Heat and Mass Transfer", Global Digital Press, Columbia, (2010).
18. M. Bahrami, "Transient Conduction Heat Transfer", ENSC 388 (F09).
19. Professor M. Zenouzi, "Transient Response Characteristics and Lumped System Analysis of Geometrically Similar Objects" October 1, (2009).
20. R. Shankar Subramanian, ""Unsteady Heat Transfer: Lumped Thermal Capacity Model", Department of Chemical and Biomolecular Engineering, Clarkson University.
21. P. K. Nag, "Heat and Mass Transfer", Second Edition, Tata McGraw Hill Publishing Company Limited, New Delhi, (2007).
22. Yunus A. Cengel, and Afshin J. Ghajar, "Heat and Mass Transfer: Fundamentals and Applications", Fourth Edition, McGraw Hill, (2011).
23. M. Thirumaleshwar, "Fundamentals of Heat and Mass Transfer", Second Impression, Published by Dorling Kindersley, India, (2009).
24. World Heritage Encyclopedia, "Lumped System Analysis", Published by World Heritage Encyclopedia.
25. Balaram Kundu, Pramod A. Wankhade, "Analytical Temperature Distribution on a Turbine Blade Subjected to Combined Convection and Radiation Environment", Journal of Thermal Engineering, Vol. 2, No. 1, January (2016), PP. (524 – 528).

26. Sepideh Sayar, "Heat Transfer During Melting and Solidification in Heterogeneous Materials", Thesis in Master of Science in Mechanical Engineering, Blacksburg, Virginia, December (2000).
27. M. Khosravy, "Transient Conduction", Islamic Azad University, Karaj Branch.
28. Lienhard, John H., "A Heat Transfer Textbook", Cambridge, Massachusetts, Phlogiston Press, ISBN 978 – 0 – 9713835 – 3 – 1, (2008).
29. Welty James R. and Wicks Charles E., "Fundamentals of momentum, heat and mass transfer", 2nd edition, New York, Wiley, ISBN 978 – 0 – 471 – 93354 – 0, (1976).
30. Paul A. Tipler and Gene Mosca, "Physics for Scientists and Engineers", 6th edition, New York, Worth Publishers, ISBN 1 – 4292 – 0132 – 0, (2008), PP. (666 – 670).

Appendix

Mathematical Formulae Summary

1. Conduction of heat in unsteady state refers to the transient conditions where in heat flow and the temperature distribution at any point of the system vary continuously with time.

2. The process in which the internal resistance is assumed negligible in comparison with its surface resistance is called the Newtonian heating or cooling process.

$$\frac{\theta}{\theta_o} = \frac{T(t) - T_\infty}{T_o - T_\infty} = e^{\frac{-hA_s}{\rho V c_p} \tau} \quad (i)$$

Where,

ρ = density of the solid, kg/m^3

V = volume of the body, m^3

c_p = specific heat of the body, $J/kg °C$ or $J/kg\ K$

h = heat transfer coefficient of the surface, $w/m^{2°}C$ or $w/m^2 K$

A_s = surface area of the body, m^2

$T(t)$ = temperature of the body at any time, $°c$

T_∞ = ambient temperature, $°c$

τ = time, s

Biot number,

$$Bi = \frac{hL_c}{k}$$

Fourier number,

$$Fo = \frac{\alpha \tau}{L_c^2}$$

Where,

L_c = characteristic length, or characteristic linear dimension.

$\alpha = \left[\frac{k}{\rho c_p}\right]$ = thermal diffusivity of the solid.

$$\frac{\theta}{\theta_o} = \frac{T(t) - T_\infty}{T_o - T_\infty} = e^{-Bi \times Fo} \quad (ii)$$

Instantaneous heat flow rate:
$$q'(\tau) = -hA_s(T_o - T_\infty)e^{-Bi \times Fo} \quad (iii)$$

Total or cumulative heat transfer:
$$Q(t) = \rho V\, c_P (T_o - T_\infty)\left[e^{-Bi \times Fo} - 1\right] \quad (iv)$$

3. Time constant and response of temperature measuring instruments:

The quantity $\frac{\rho V\, c_P}{hA_s}$ is called time constant (τ^*).

$$\frac{\theta}{\theta_o} = \frac{T(t) - T_\infty}{T_o - T_\infty} = e^{-(\tau/\tau^*)}$$

The time required by a thermocouple to reach its 63.2% of the value of the initial temperature difference is called its sensitivity.

4. Transient heat conduction in semi – infinite solids (h or $Bi \to \infty$): The temperature distribution at any time τ at a plane parallel to and at a distance x from the surface is given by:

$$\frac{T(t) - T_\infty}{T_o - T_\infty} = erf\left[\frac{x}{2\sqrt{\alpha\, \tau}}\right] \quad (i)$$

Where $erf\left[\frac{x}{2\sqrt{\alpha\, \tau}}\right]$ is known as "Gaussian error function".

The instantaneous heat flow rate at a given x – location within the semi – infinite body at a specified time is given by:

$$Q_i = -kA(T_o - T_\infty)\frac{e^{[-x^2/4\alpha\tau]}}{\sqrt{\pi\, \alpha\, \tau}} \quad (ii)$$

The heat flow rate at the surface ($x = 0$) is given by:

$$Q_{surface} = \frac{-kA(T_o - T_\infty)}{\sqrt{\pi\, \alpha\, \tau}} \quad (iii)$$

The heat flow rate $Q(t)$ is given by:

$$Q(t) = -1.13 kA(T_o - T_\infty)\sqrt{\frac{\tau}{\alpha}} \quad (iv)$$

About the Author

Osama Mohammed Elmardi Suleiman was born in Atbara, Sudan in 1966. He received his diploma degree in mechanical engineering from Mechanical Engineering College, Atbara, Sudan in 1990. He also received a bachelor degree in mechanical engineering from Sudan University of Science and Technology – Faculty of Engineering in 1998, and a master degree in solid mechanics from Nile Valley University (Atbara, Sudan) in 2003. He contributed in teaching some subjects in other universities such as Red Sea University (Port Sudan, Sudan), Kordofan University (Obayied, Sudan), Sudan University of Science and Technology (Khartoum, Sudan) and Blue Nile University (Damazin, Sudan). In addition, he supervised more than hundred and fifty under graduate studies in diploma and B.Sc. levels and about fifteen master theses. He is currently an assistant professor in department of mechanical engineering, Faculty of Engineering and Technology, Nile Valley University. His research interest and favorite subjects include structural mechanics, applied mechanics, control engineering and instrumentation, computer aided design, design of mechanical elements, fluid mechanics and dynamics, heat and mass transfer and hydraulic machinery. He also works as a consultant and technical manager of Al – Kamali workshops group for small industries in Atbara old and new industrial areas.